The Ideal Soil v2.0
A Handbook for the New Agri

This Book is Respectfully Dedicate
William A. Albrech

The Ideal Soil method evolved from the research of Dr William Albrecht and his associates at the University of Missouri Agricultural Experiment Station in Columbia, Missouri from the 1920s through the early1960s.

William A. Albrecht (1888–1974) PhD, Chairman of the Department of Soils at the University of Missouri 1938-1959. President of the Soil Science Society of America 1939.
The world is changed by those who show up for the job.

The Ideal Soil v2.0: A Handbook for The New Agriculture
By Michael Astera with Agricola

ISBN # 978-0-9844876-9-1 Hardcopy
Cover photo *Mineral Wheel.* Concept, design, and photography by Elizabeth Brown, RoseMinerals.com

Table of Contents

Appendices continued:

Foreward to The Ideal Soil v2.0

Since the first e-book version was published in December of 2008, The Ideal Soil Handbook has gone around the world, and the Ideal Soil method has been applied in every imaginable climate and soil type. From Australia to Japan, South Africa to Finland, from Argentina to Alaska, and almost everywhere in between, readers have balanced the minerals of farms, ranches, greenhouses and backyard gardens. The method is in use on coffee plantations in the highlands of Laos and the hills of Zambia, rice farms in Thailand and the Philippines, horse pastures in Borneo and sheep pastures in Oregon. As the 2015 English edition goes to print, the first Spanish edition is being printed along with it, with a Dutch translation almost complete and the beginnings of an Indonesian version. At least two recently published books in English were inspired by and are based on this book. In short the Ideal Soil book has been well received and the method has been proven successful wherever it has been applied. Version 2.0 is the result of what we have learned. Except for the introductory chapter and parts of the appendices, it has been completely rewritten and much new material has been added.

The two most important additions are Chapter 9 on Calcareous and High-pH Soils, and Chapter 10 on Working with Low-CEC soils. The biggest problem encountered was how to get an accurate estimate of CEC in order to balance the minerals in high pH soils; that has been solved with the Ammonium Acetate pH 8.2 test, explained in Chapter 9. The second biggest problem was what to do with low CEC soils, below 7meq. Chapter 10 explains the minimum amounts needed for the Ideal Soil method to work and suggests numerous ways to increase CEC.

The other major change is that the method is explained using parts per million, ppm. When the amounts are calculated this way, it is simple to convert the results to pounds per acre, kilograms per hectare, grams per cubic meter, or any other system of weights and measures.

We now have data showing the increased nutritional value of crops grown in mineral balanced soil. Once nutritional superiority is demonstrated, proven, that will not just set the bar higher, it will set a whole new bar. The world of agriculture needs a new direction, and this is the logical next step for a technological society that cares about a healthy world. At the same time, the new agriculture isn't all hard science and logic. The wisdom of the heart and the knowingness of intuition are not to be neglected. Sincere thanks to all of you who are contributing new knowledge and insights, and helping to spread the vision. Not only can we do this, together we are doing it.

Agricola
19th November 2015

Preface to the First Edition (2008): Towards a New Agriculture

This is a book about the New Agriculture, but more than being only *about* the New Agriculture, it is a how-to manual that tells exactly how to go about changing your farming, gardening, and landscaping practices over to a sustainable and healthy model that does not require large year to year inputs of fertilizers or massive inputs of organic matter. Once the soil minerals are balanced and part of a living soil, the need for insecticides and other "rescue chemicals" will drop away as well. The New Agriculture is premised on the idea that being well fed leads to health. Living things that are well fed do not attract disease or parasites. From the smallest bacterium to the largest tree all living things have a genetic potential for growth and health that is only limited when something is missing from or out of balance in their environment. As gardeners, farmers, and caretakers of the land we cannot control the variables of climate and we cannot control rainfall; what we can control is the level and balance of essential nutrients in the soil, but that is plenty.

A solid understanding of minerals has been the weakest or even missing leg of sustainable agriculture up until now. The purpose of this book is to put into the reader's hands the tools needed to apply our present knowledge of soil minerals in a practical way, not just about why minerals are necessary or what they do, but exactly how to apply that knowledge: How much of what mineral in what form? That is the question that we came to soon after starting to study the subject and the one that we could not find the answer to in any of the other books or on the world wide web. How much Potassium, how much Phosphorus, how much Zinc? In these pages you will find the answers, answers that have been tested and proven to be highly effective and safe.

In order to begin, we need first of all to know what we are starting with, we need to take inventory. This book is a recipe for creating the ideal soil, but that recipe will do no good without knowing what ingredients we have to start with, so that we know what other ingredients may be lacking. The only way to know what we are starting with is to have the soil assayed by a soil testing laboratory. Once that test is in hand, the rest is pretty simple. Without the soil test results, we are floundering in the dark, we are merely guessing. Soil testing is not expensive. There are many soil testing labs around; they have hundreds of thousand of dollars of precision scientific equipment and skilled technicians to operate it, and they are inexpensive and fast. You will need the results of a soil test to use the information in this book.

We have held back nothing that we know. It's all here, everything that we have learned over many years of reading, research, and experimenting. To the best of our knowledge, no book like this has ever before been published; no one before this has tied all of the major and minor soil nutrients together and shown their correct proportions and relationships to each other. Our fond hope is that those who read and understand The Ideal Soil will no longer need us, but will have all of the information to go out and bring their own soil to their own version of perfection, and perhaps go on from there to offer that service to others.

At present there are very few persons with the knowledge of how to balance the mineral nutrients in the soil; perhaps a few dozen competent consultants on the subject in the USA, and not many more worldwide. None of them before now have had access to the system explained in this book. Until this first on-line publication in November of 2008, this information has been our trade secret, known to fewer than six people. We have used the Ideal Soil chart proportions for a number of years in many different climates and soils, and can confidently state that we have had few if any problems and no complaints. It just works.

For the New Agriculture to come into its own will require many more people to understand these principles. Perhaps one day each community will have a trusted soil physician just as they once had a trusted family physician. Perhaps you the reader will be one of them. We sincerely hope so.

The science of soil nutrients can trump remedial nutrition and it can trump pharmaceutical medicine. It is primary; it comes before either of the above and if applied intelligently can make them unnecessary. A healthy well fed body will suffer from no nutritional deficiencies and will need no supplements or drugs.

There is nothing difficult about learning to balance the soil minerals. A very rudimentary understanding of chemistry and fifth-grade arithmetic are the only bits of knowledge needed. Every step in the process and the reason for doing it are clearly explained and shown. You are invited to take this knowledge, put it into practice, add to it, and make it your own. You can do this.

Agricola
Terra, November 6, 2008

Chapter 1
The New Agriculture: What It Is, and What It Is Not

How did we end up where we are with our food supply today? Most would admit it looks pretty grim. Setting aside looming food shortages and price inflation worldwide, how did we end up with such abysmal nutritional quality coupled with high levels of noxious chemicals and compounded by deteriorating agricultural soils around the world?

Unsurprisingly, it has the same roots as our present abysmal economic prospects, being rooted in short-sighted greed coupled with ignorance and manipulated for the benefit of a few at the expense of the rest. Unlike economics, however, who stands to gain when the whole of humanity is ill and malnourished? Not humanity, that's for sure.

The wealthy may have more money and more security, but their food is no better than that of the average peasant and often worse. The falling tide of nutritional quality in food has left everyone's boat high and dry. Surely the wealthy aren't starving for bulk of food, but they suffer from the same diseases of malnutrition and toxicity as the rest of us do, namely cancer, diabetes, heart disease, and the various auto-immune diseases ranging from MS to AIDS. Whether dining in the fanciest restaurant or the poorest hut, nutrient deficiency and toxic overload are on everyone's plate.

This is the situation bequeathed to us by a century and a half of the increasing dominance of agriculture by a corporate industrial model focused solely on yield and profit. The truth of these observations is undeniable to anyone who looks objectively at world agriculture today. There are other schools of agriculture that have rejected the chemical industrial model and deserve great credit for their struggle to grow clean food and create a healthy environment in harmony with Nature.

On the following pages we will take a look at where, in our opinion, the alternatives too fall short of the goal of being truly sustainable or providing the best possible food. We will also learn a little of the history of mineral balanced agriculture and its present role in world food production. None of the following is meant to offend, but it is not sugar coated.

What The New Agriculture Is Not

All of today's agriculture movements clamor that they have the answers, but do they? This writer thinks not.

The "better living through chemistry" factions are still flogging their tired horse. Having stripped the soil of its richness, burned out the humus and killed off the soil life, and having turned much of their not-so-little corner of Nature into a nutrient depleted toxic wasteland, they are now developing Frankenstein's monster crops, genetically modified organisms or GMOs, bred to live in these conditions. We can count on this turning out as well as their previous bright ideas.

This book is all about science and chemistry, but science and chemistry in the service of humanity and in harmony with Nature, not science and chemistry misused in a vain attempt to exploit and beat Nature into submission.

Humans are a self-aware and intelligent land animal. We have eyes and ears and brains; legs and arms and hands with opposable thumbs. We have the ability to understand the present and envision the future. Our role should be that of caretakers of our home, as we are the only ones who can do that. An intelligent person does not cut down the tree that shades their house from the hot afternoon sun or pour sewage in their family's drinking water, Attempting to exploit our only home for short term gain makes no logical sense; obviously it hasn't worked, isn't working, and won't work in the future.

The worldwide Organic agriculture movement and its various offshoots have so far only offered simplistic solutions, mostly one simplistic solution: add more organic matter to the soil. This is the school from which this book's authors come, and most growers with whom we work are organic growers. “More organic matter” is a step in the right direction if the soil is low in humus, but does little to address nutritional deficiencies, especially mineral deficiencies. Yet it is fiercely defended and proclaimed to be “the answer” for everyone everywhere. Is it? No. While essential, soil biology and organic matter are only a part of what makes a healthy soil and nutrient dense crops. Nature is not simple, and simplistic one-size-fits-all answers are not going to solve the nutritional and environmental crises we face.

Those who follow the Biodynamic school are to be commended for their deep appreciation of Nature and for having preserved much traditional knowledge and brought it into the present. They have an understanding of energy that goes far beyond simple electrical current flow, but by not fully understanding the minerals in their soil, they limit their potential.

Permaculture works fine in many instances, but is mostly an approach to stabilizing the existing soil, preventing erosion. Under a permaculture system the nutrients that are in the soil are largely retained; what is taken away is supposedly replaced with a fresh layer of organic matter. If every bit of the crop that was taken away was somehow brought back and replaced, the soil nutrient content would still only be what it was to start with, which in the case of most agricultural soils is far from ideal.

The various fans and promoters of soil biology, from earthworms to fungus, tell us that a bio-active soil will break down toxic residues, increase humus, and the beneficial soil organisms will make minerals and nutrients available to the plant. The question that is not asked is “what if the needed minerals are not to be found in the soil?”

The newer high-tech solutions, such as hydroponics, or even newer, aeroponics, rate a careful examination. Can we count on them to rescue agriculture? Not if the goal is to feed the world's people and animals. They are fine for growing some pretty tomatoes to sell at the supermarket, or some nice lettuce in the basement, but these "new and modern" systems have a number of basic problems, some of them insurmountable if the goals are sustainability and nutrient-dense food. The most obvious failing is that they are energy-hungry. They use pumps and fans and often lights. In the interests of self-sufficiency, where is that energy to come from? If the power goes out is one going to pedal a bicycle generator to keep the pumps and fans going? In addition to being energy-hungry, both hydroponics and aeroponics require special containers, growing solutions, training and handling. They are not automatic.

There are other not so obvious problems with hydroponics. Any time one has a liquid-based growing solution they need water-soluble fertilizers, and these must be pure. One does not put compost in the hydroponic trays. This makes all natural organic hydroponics pretty difficult. Another drawback is that only certain crops are suitable, mostly the ones you have seen in the stores so far: lettuce, tomatoes, peppers, and some herbs. One will not raise a field of potatoes, cassava, or turnips hydroponically, nor thousands of acres of grains and legumes. One will not grow hay to feed animals hydroponically or aeroponically.

The most serious downside to these systems, though, is the lack of nutritional completeness in the produce. Designer vegetables grown in nutrient solutions are grown for looks, not nutrition. No one has yet shown that a nutritionally complete diet can be grown in this artificial manner.

Mention should be made of the ultimate closed-environment theory of the day (or decade), the all-in-one fish pond and hydroponic garden. As you may know, the idea is that one raises fish in a pond, then uses the fish water to irrigate the hydroponic troughs. The nutrients from the fish water are used as fertilizer for the plants. The water comes out "clean" at the other end and is recycled back to the fish pond. Various theories suggest what the fish eat, but the grower gets to eat the fish and the vegetables. The theory sounds good, but all the designs seem to require glass or plastic domes. We will not feed ourselves and heal our polluted environment by creating isolated bubbles in the landscape.

The high-tech systems above are things to learn from and we will and have gained knowledge from them. One valuable contribution is that we know more about what mineral nutrients are absolutely essential for plant growth. These systems, however, are not suitable for feeding your family and community, and they will not form the basis of the New Agriculture.

The place to grow a crop is in the earth, in nutrient rich, biologically active soil, not in metered nutrient solutions; under natural sunlight, not electric lights. Sunlight is very energy-dense and plants are good at using it. Sunlight is also free. Within the limits of one's climate, one can create micro-environments that maximize solar gain, and one can choose crops that do well under one's local conditions. In Alaska and Finland, one might choose to grow cabbages, not melons.

The New Agriculture will not come about through dogmatic insistence on simplistic solutions such as adding organic matter to the soil, nor through force-feeding of synthetic fertilizers and applying toxic rescue chemicals to address the inevitable problems. The answers will not be found in energy intensive technology or artificial micro-environments. The solutions certainly won't be found by refusing to look outside whatever ideological box one has adopted or been convinced to adopt.

The New Agriculture

What we have today is a fragmented agriculture, yet we needn't be suffering this collective delusion and separation; it serves no useful purpose for mankind or Nature but only divides us. So here's a proposal: What if we were to take agriculture to another level, a higher level, by pulling together the best from all of modern knowledge, and combining it with the traditional wisdom accumulated over the span of human history? What if we were to include the sciences of soil chemistry and nutrition (new tools in the 10,000 year history of agriculture), with a modern understanding of soil and plant biology (also new tools), and our modern knowledge of energy, both electromagnetic and subtle? The only questions we need ask are: What works and will continue to work, and what hasn't worked in the past or doesn't work now? No special emphasis would be laid on any one dogma or school of agriculture; the focus would be on soil health, nutrition, sustainability, and efficiency. The emphasis would be on constant improvement in health: of the land, the plants, the animals, and the people.

We would be looking for a system that works well with any crop in any climate, producing high yield, high quality, and high nutritional values while sharply reducing insect and disease problems. The plants would thrive and be superbly healthy because they would have all of the nutrients they desire available free-choice. The immune systems of the plants and soil would be strong and healthy; insects and disease are not attracted to strong, healthy plants. The animals and people consuming the plants would get the most highly nutritious food it was possible to grow. People wouldn't overeat because their body wouldn't be craving an essential mineral, carbohydrate, amino acid, or lipid. Diseases such as diabetes, cancer, heart disease, and the auto-immune diseases would become things of the past. Children would grow up able to develop to their full genetic potential; their intelligence and strength would no longer be limited by malnutrition or toxic chemicals. Fewer acres of cropland could feed more people and animals, sustainably, as the emphasis shifted from quantity to quality.

Unbeknownst to most, the basis of this new agriculture already exists and has for some time. The knowledge of how to accomplish the goals mentioned above has largely been known for over sixty years. The basic science of soil mineral balance and its relation to health and nutrition was discovered long ago, but has been buried and ignored. It has been hidden from the schools and practitioners of agriculture, both so-called "conventional" and the various alternative schools. It is not mentioned, or mentioned disparagingly in university ag colleges. Many "alternative" growers have never heard of it. Those who have heard of it but don't understand it and have never tried or experienced it nevertheless have opinions on why it couldn't work. We are in the situation of having the answers readily available but blindly refusing to see them.

Much of the work this mineral balanced agriculture relies on was done in the 1920s, '30s. and '40's. During the depression era of the 1930s there was a strong emphasis on finding out what went wrong in agriculture that led to the dust bowl years and a general decline in the health of American soils and people. Scientific nutrition was a new field and many exciting breakthroughs were made. By the late 1930s and early 1940s great strides were being made in both soil and animal health.

Along came WWII, and the food producers (farmers) were urgently needed; they were recruited by the government and made part of the war machine, subsidized by guaranteed crop prices, and were encouraged to innovate. The end of WWII saw most of the economies of the industrialized world dominated by the factory production model, much of it war-related. After WWII this industrial model was re-directed into the production of goods, machinery and chemicals for peacetime.

By 1950 it appeared to be a brave new modern world, one where all problems could be solved by dominating Nature, rather than learning from and cooperating with her. Big chemical companies took over the land grant universities and started really pushing their chemical-based agriculture. Most of the farmers eagerly adopted the new model; no longer were they just farmers, they were modernized commodity factories on the cutting edge of science. Or so they thought. While the yield went up, the nutritive value fell, and the plants force-grown on soon-depleted soil were insect and disease magnets, calling for more chemicals every year. The harsh concentrated fertilizers burned up the humus in the soil and killed off soil life. The soils were robbed of their mineral stores, as the only nutrients applied were those necessary to achieve high yield. The animals (and people) raised on these force-fed foods became malnourished and disease-prone. The law of diminishing returns was showing up with a vengeance, but the "scientific" solution to the problem was always another and more powerful chemical and a plant bred to tolerate it.

Meanwhile, still shortly after WWII, J.I. Rodale started the organic gardening movement in the USA, inspired by the work of Sir Albert Howard and Lady Eve Balfour in the UK, while William Albrecht was proving the validity and value of mineral balanced agriculture. William who?

The late William A. Albrecht, PhD, and his crew of researchers at the University of Missouri agricultural station were responsible for developing the mineral basis of the New Agriculture: the concept of balancing the alkaline nutrients in the soil based on the soil's capacity to hold them. In the 1920s they decided to take a close look at the various mineral fractions of soil: the clay, silt, and sand fractions. They took some of the local soil, removed the organic matter, and spun it in a centrifuge to separate it by size and weight. This yielded an almost clear, jelly-like layer on top that turned out to be made up of incredibly tiny clay particles, particles too small to be viewed by most microscopes. They were so tiny that they stayed suspended in water and wouldn't even centrifuge out, though they didn't dissolve. Colloids are what this type of particle is called; this was colloidal clay. What did these tiny bits do in the soil? It turned out they did a lot. Those colloidal clay particles were the basis of the soil's cation exchange capacity. They stored the alkaline nutrients in the soil, held by a simple static electrical charge, safe from being washed away, yet readily available to soil life. The plants and soil life traded +charged Hydrogen ions for these + charged nutrients. Albrecht and crew spent the next three decades

experimenting with various combinations of mineral nutrients, growing the crops and feeding them to animals, measuring the nutritional value of the crop and the health of the animals.

However, by the late 1950s and early '60s the big chemical companies had managed to take over most of the USA's agricultural schools. They offered to fund new buildings and research projects, and pay for new professorial chairs, but Professor Albrecht and the other holistic researchers from the 1920s, '30s, and '40s had to go. Albrecht had demonstrated that the chemical companies' approach was an unnecessary path to bankruptcy and destruction and he wasn't about to teach their party line, especially as he had developed and spent years proving a better system that was sustainable and healthy. Albrecht was forced into retirement in the 1960s; his work was buried and would have been lost if not for the efforts of economist and editor Charles Walters, who started the magazine AcresUSA in 1970 to promote Albrecht's ideas. Charles Walters called this new science of balancing the cation minerals in the soil Eco-agriculture. It has been implemented on hundreds of thousands of acres of commercial farms in the US and Australia with great success, but the mineral balancing message hasn't yet gotten to the home gardener or small producer, nor has it gotten to the various branches of alternative agriculture. The corporate-dominated State Agriculture Colleges pretend it doesn't exist.

J.I. Rodale worked with Wm Albrecht and Louis Bromfield at Bromfield's Malabar Farm in Ohio during the late 1940s. Bromfield was working to restore worn-out farmland by applying Albrecht's mineral balancing principals as well as the organic ideas of the English agriculturist Sir Albert Howard. The story is that Rodale had a falling out with the Malabar farm group over the use of some man-made fertilizers that the others considered not to be harmful, probably ammonium sulfate. Rodale was a purist and his version of organic had no room for input that wasn't 100% natural. Sir Albert Howard taught that trees and other deep-rooted plants would bring up any minerals needed, and didn't give it a lot of thought beyond that. Rodale was convinced that leaves from deep-rooted trees, and rotting vegetable matter in general, could supply all of the nutrients plants needed to thrive, even in poor or worn out soil.

Rodale went on to found Organic Farming and Gardening magazine, today's Organic Gardening magazine, and for the first ten years almost all he wrote about was organic matter; mulch and compost were all anyone needed, he seemed to think. Only later, starting in the 1960s, did he begin to acknowledge the role of minerals and recommend them, particularly rock phosphate, greensand, and dolomite lime; but ordinary garden lime, Calcium, was seen merely as a pH adjuster, instead of being recognized as the single nutrient needed in most quantity in the soil that it actually is. J.I. Rodale was a man with a mission, and all of us who learned from him owe him great honor. He was almost single-handedly responsible for inspiring the strong and vibrant organic agriculture movement in the USA and around the world today. Anyone whose education in gardening was in the Rodale school, however, is going to know that minerals are needed, but is unlikely to know why or how much or where they come from.

Meanwhile, Albrecht's mineral balanced agriculture, as promoted by Walters in the AcresUSA newspaper and a number of books, moved forward through the 1980s and '90s, but only on good-sized farms, and few enough of them. Very few of the farmers using the mineral approach knew much if anything about the organic crowd. Balancing soil nutrients based on the soil's exchange capacity worked and worked well, and when a farmer had had enough of chemicals and poisons, or saw his neighbor growing better crops than he while working less and spending less, many did apply Professor Albrecht's principles and they continue to do so today. I have heard of no one switching back to their earlier style of farming, gardening, or ranching once they have experienced the results of a mineralized, balanced soil.

Another important person in bringing the knowledge of mineral nutrition to agriculture was the late Carey Reams, PhD, who did most of his life's work in Florida, USA. The Albrecht and Reams schools have slightly different but easily reconciled philosophies; they agree on the mineral balance, but often use different explanations and terms. Students of Carey Reams and Wm. A. Albrecht, and the students of their students, make up most of the mineral-aware agricultural consultants around today, worldwide, including this author.

Organic gardening, unfortunately, was stuck back in the 1950s, and it has largely remained there since: Compost, manure, mulch, and that's about it. The other schools of alternative agriculture - Steiner's Biodynamics, Permaculture, Elaine Ingham's Soil Food Web concept, the various miracle microbe schools etc.- all emphasize the biological and compost-based approach almost exclusively. The occasional mention is made of rock dust, phosphate rock, or dolomite lime, but seldom with any understanding of the soil chemistry involved.

The one truly mineral-oriented school of "mainstream" alternative agriculture is what I call the Glacial Rock Dust school, based on the famous 1982 book *The Survival of Civilization*, whose authors argued that the retreat of the glaciers at the end of the last ice age was the last time our soils had a fresh dose of minerals. Their solution was to add freshly ground rock powder to the soil as the source of those missing minerals, but there is little understanding of the actual role of minerals, and no conception of the amounts or balance of minerals needed. A average everyday soil with a cation exchange capacity of 10 requires around 3,000 lbs of Calcium in exchangeable form per acre, and 50 or so pounds of Zinc. Is that in the rock dust or not? Does the soil need the minerals in that particular rock dust at all? Freshly ground rock dust is a great soil amendment, but it can't be counted on to correct a mineral imbalance or deficiency.

What the USA ended up with by the 1970s was a great division between those practicing organic agriculture and those farming with strong, concentrated chemical fertilizers, pesticides and herbicides. Neither side talked to the other, the organic group taking the moral high ground against poisoning the land and the chemical farmers deriding the organic followers as backwards Luddites. Neither side knew about the successes of those using the methods of Albrecht or Reams. How could they? Organic Gardening was heavily invested in the idea that organic matter and soil biology alone were the answers, while the chemical farmers were convinced that the next hybrid crop and the newest pesticide were going to solve their growing problems. Neither one was interested in learning that they were both wrong, that there was a system already up and running that didn't require scores of tons of compost and manure per acre and didn't need toxic rescue chemistry either.

Our Story Continues Today

Back at the corporate laboratories and bought-off State agriculture colleges, the dyed-in-the-wool chemical farming fans are still trying to prove that the growing of food can be forced into an industrial production model. Their version of "working with biology" up until the 1990s was hybrid crops, and has now morphed into GMOs, genetically modified organisms. Both the hybrids and the GMOs are usually plants that have been bred to live on a starvation diet of NPK fertilizer while being regularly doused with herbicides, fungicides, and insecticides. Yield, disease resistance, the ability to survive repeated dosing with noxious poisons---these are the goals of the mad scientists leading corporate chemical agriculture. The health of the soil and the nutritional value of the crop are meaningless to them. Is this too harsh a judgment? Look at the nutritional quality of our food and the worn-out state of our farmlands to answer that question

I'd like to insert a rather esoteric opinion here. It is my contention that attempting to turn agriculture into an industrial process breaks a fundamental agreement that mankind has had with nature since the inception of thinking humans on this planet. Not only with nature in general, but with the individual plant and animal families with whom we have these ancient agreements. The agreement with cattle, for instance, is that their human herders will offer protection from wild predators, shelter and warmth when necessary, and provide good food and water to them. We will help protect their offspring, care for them when they are sick or injured, and work to improve the breed. In exchange, the cattle provide for us their milk, meat, hides, manure, and sometimes labor. This has been a fair trade for the animals and for the humans taking on the responsibility.

We humans have long had a similar agreement with members of the plant kingdom: care, protection from competing plants, fertile soil and abundant water, working to improve the breed. Industrial agriculture and corporate greed have broken these agreements, and more than broken them: these ancient pacts have been violated in the most obscene manner. An old English term for a farmer and livestock person was a husbandman. To husband was a verb that meant to care for as a wife's husband would care for their family: To husband the land, and the crops, and the animals. Wise husbandmen passed on a better farm than they inherited, passed this on to their children and to the descendants of the plants and animals they had cared for and partnered with. We who wish to create a new and better world should strive to get back to that ideal, and to extend it to all of the Earth that is in our care.

Getting back to our critique of today's agriculture: Regardless of their intent, neither the granola heads nor the nature nazis have proven to have much of a clue when it comes to the big picture. It's time to change that situation. In order to make a new agriculture, we need to use everything we know or can find out, from any discipline. Being a believer and purist of any one school or philosophy of agriculture, and trying to bend reality to fit those accepted truths, is not going to lead us forward.

Most organic growers have no clue what minerals are in their soil. Is it not so? The chemical growers are generally a little better informed, as they are used to getting their soil tested in order to find out how many pounds of chemical fertilizer to add, but they have little understanding of the essential role of the nutrient minerals either.

Our physical reality is made of minerals, also known as elements. There are 90 or so naturally occurring elements, from Hydrogen to Uranium, and we don't really know how many of them we need in order to live, but it's a lot of them. We must have Iron to transport Oxygen in the blood. Calcium and Phosphorus are used to build the crystal lattice of our bones and teeth. Lack of Zinc causes sterility, decreased brain development, loss of sensory acuteness. When the immune system is threatened by infection it releases its stores of Copper from the liver and pulls Iron from the blood. Many metals are re-used over and over as catalysts in the formation of proteins and amino acids. They serve as templates, shapes, that the proteins are folded around. The shape of the protein determines its fit into its intended destination in a living cell. The health, growth, and reproduction of all living things is dependent on the availability and proper balance of mineral elements.

Despite the pervasive ignorance in agriculture, we all know from our nutritional knowledge that minerals are essential to our health. How many people take a vitamin/mineral supplement? Or Calcium supplements? The science of nutrition is well aware of essential minerals, and nutrition books, radio programs, and websites are always decrying the lack of minerals in our food, telling us how the soil is depleted of minerals, and how we can save ourselves from this menace by taking a mineral supplement. Meanwhile, the organic food promoters keep claiming that organically grown food has more minerals, without having a clue whether that's true or not, and in most cases without having an inkling if there are actually any minerals in the soil at all.

Why the disconnect? If minerals are not in the food it's because they are not available in the soil. So why not add them to the soil and get them in your food? At the same time, feed and activate the soil life, bring the humus level up to optimum for your soil and climate, and provide the energy the plants and soil life need. The soil will be healthy, the plants too, and so will the people and animals who eat the nutrient-dense food grown in the Ideal Soil.

If we look at agricultural soils from a nutritional standpoint, they are much more than an anchor for the roots, a base to keep the crops from falling over. Each crop harvested and taken away depletes the soil's store of essential nutrient minerals. If the minerals are not replaced, we eventually reach a point where there are not enough left to grow a healthy crop with the ability to mature seeds for the next generation. Long before this point is reached, the nutrient density of the crop for human and animal food has suffered. Much of our arable land worldwide is producing empty calories, mostly carbohydrates made from the atmospheric elements Carbon, Hydrogen, and Oxygen. The solution, the only solution (barring the ability of plants or soil organisms to transmute elements alchemically), is to supply these needed minerals from a source where they are abundant. That source should ideally be located as close as possible to where the minerals are needed in order to minimize transportation costs. It makes no sense to ship ground limestone across the country when every state in the USA has limestone deposits, but when it comes to rare elements like Selenium or Boron which are only found in concentrated form a few places in the world, the transport costs are justified.

Mining of the needed minerals need not entail long-term environmental damage either. Mines and quarries can be carefully worked by those who care about their home planet, and when the mines are depleted they can be landscaped and planted to be as or more beautiful than before mining. It's also worth noting that many of the economically viable sources for agricultural minerals contain such high concentrations of these minerals that they are toxic to soil life and little or nothing grows there. Removing these toxic concentrations and using them to make other parts of the planet healthier and more productive can, at the same time, open up these formerly toxic soils to the growth of forest or grasslands. None of this should be done on the basis of greed or short-term gain, but rather wisely, intelligently, and in harmony with Nature.

A wonderful thing about a balanced, mineralized soil based on the soil's exchange capacity is that everything else becomes easier. The soil pH self-adjusts to its optimum, plant disease and insect problems largely disappear, water retention, drainage, soil texture, and rate of decay of organic matter all become self-regulating and automatic, weather permitting. The grower knows that the nutrients are in the
crop because the nutrients are available in the soil. The soil life is active and healthy and helping to make these nutrients available, and the plants growing on this ideal soil have free-choice of any nutrient they want, in balance, a balance designed by intelligent science and observation.

All of this can be achieved using minerals in the form of naturally-occurring rocks and mineral ores or their purified forms, ancient sea-bed deposits, ocean water, and the byproducts from plants and animals. The cultural practices one is presently using may change little, except to become easier. This is real science in harmony with nature, using all of the best of ancient and modern knowledge intelligently: the New Agriculture.

There are a few simple and basic principles that govern soil mineral balance. The most important to understand is the soil's Cation Exchange Capacity, or CEC, often referred to simply as exchange capacity or EC. This is a measure of the quantity of nutrients and non-nutrients the soil can hold, how big its "holding tank" is. The lower the tank gets, the more the soil life and plants have to struggle to get their nutrients. On the other hand, if one applies more nutrients than the soil can hold, those nutrients will wash away in rain or irrigation water, or build up in the soil. Excess nutrients are either unnecessary or harmful. One would not put 30 gallons of gasoline in a twenty gallon tank and expect to gain anything. Exchange Capacity EC is the amount the soil can hold onto and use. One must know their soil's exchange capacity, and its % of saturation by different nutrients, to know where one is now, and where one needs to go. In the next chapter we will gain a working understanding of the soil's cation exchange capacity.

If the songbirds are singing, we are getting close.

The Ideal Soil: A Handbook for the New Agriculture

Chapter 2

Cation Exchange Capacity in Soils, Simplified

(Revised December 2013)

Adsorb vs Absorb

adsorb(ad sôrb, -zôrb), v.t. Physical Chem. to gather (a gas, liquid, or dissolved substance) on a surface in a condensed layer: Charcoal will adsorb gases .

Please note the definition above, taken from the large hardbound version of the Random House Second Edition Unabridged Dictionary. It's not absorb, it's adsorb , with a "d". We all know that a sponge absorbs water, a cast iron pot absorbs heat, a flat-black wall absorbs light. None of those gathers anything on the surface in a condensed layer, they soak it right in, they absorb it.

Adsorb is different, because it means to gather on a surface in a condensed layer. This is pretty much the same thing as static cling, like when you take a synthetic fabric shirt out of the clothes dryer and it wants to stick to you. You don't absorb the nylon blouse, you adsorb it. Everyone got that? Good. On to Cation Exchange Capacity.

The Exchange Capacity of your soil is a measure of its ability to hold and release various elements and compounds. We are mostly concerned with the soil's ability to hold and release plant nutrients, obviously. Specifically here today, we are concerned with the soil's ability to hold and release positively charged nutrients. Something that has a positive (+) charge is called a cation, pronounced cat-eye-on. If it has a negative charge (-) it is called an anion, pronounced ann-eye-on. (Both words are accented on the first syllable.) The word "ion" simply means a charged particle; a positive charge is attracted to a negative charge and vice-versa.

Positively charged particles are known as cations. There are two types of cations, acidic or acid-forming cations, and basic, or alkaline-forming cations. The Hydrogen cation H+ and the Aluminum cation Al+++ are acid-forming. Neither are plant nutrients. A soil with high levels of H+ or Al+++ is an acid soil, with a low pH.

The positively charged nutrients that we will be discussing here are Calcium, Magnesium, Potassium and Sodium. These are all alkaline cations, also called basic cations or bases. Both types of cations (alkaline or acidic) may be adsorbed onto either a clay particle or soil organic matter (SOM). All of the nutrients in the soil need to be held there somehow, or they will just wash away when you water the garden or get a good rain storm. Clay particles generally have a negative (-) charge, so they attract and hold positively (+) charged nutrients and non-nutrients. Soil organic matter (SOM or just OM) has both positive and negative charges, so it can hold on to both cations and anions.

Both the clay particles and the organic matter have negatively charged sites that attract and hold positively charged particles. Cation Exchange Capacity is the measure of how many negatively-charged sites are available in your soil.

The Cation Exchange Capacity of your soil could be likened to a bucket: some soils are like a big bucket (high CEC), some are like a small bucket (low CEC). Generally speaking, a sandy soil with little organic matter will have a very low CEC while a clay soil with a lot of organic matter (as humus) will have a high CEC. Organic matter (as humus) always has a high CEC; with clay soils, CEC depends on the type of clay.

Base Saturation %

From the 1920s to the late 1940s, a great and largely un-sung hero of agriculture, Dr. William Albrecht, did a lot of experimenting with different ratios of nutrient cations, the Calcium, Magnesium, Potassium and Sodium mentioned above. He and his associates, working at the University of Missouri Agricultural Experiment Station, came to the conclusion that the strongest, healthiest, and most nutritious crops were grown in a soil where the soil's CEC was saturated to about 65% Calcium, 15% Magnesium, 4% Potassium, and 1% to 5% Sodium. (No, they don't add to 100%; we'll get to that.) This ratio not only provided luxury levels of these nutrients to the crop and to the soil life, but also strongly affected the soil texture and pH.

The percentage of the CEC that a particular cation occupies is also known as the base saturation percentage, or percent of base saturation, so another way of describing Albrecht's ideal ratio is that you want 65% base saturation of Calcium, 15% base saturation of Magnesium etc. Don't get too hung up on these percentages; they are general guidelines and can vary quite a bit depending on soil texture and other factors.

It's still a little-known fact that the Calcium to Magnesium ratio determines how tight or loose a soil is. The more Calcium a soil has, the looser it is; the more Magnesium, the tighter it is, up to a point. Other things being equal, a high Calcium soil will have more Oxygen, drain more freely, and support more aerobic breakdown of organic matter, while a high Magnesium soil will have less Oxygen, tend to drain slowly, and organic matter will break down poorly if at all. In a soil with Magnesium higher than Calcium, organic matter may ferment and produce alcohol and even formaldehyde, both of which are preservatives. If you till up last years corn stalks and they are still shiny and green, you may have a soil with an inverted Calcium/Magnesium ratio. On the other hand, if you get the Calcium level too high, the soil may lose its beneficial granulation and structure and the excessive Calcium will interfere with the availability of other nutrients. If you get them just right for your particular soil, you can drive over the garden and not have a problem with soil compaction.

Because Calcium tends to loosen soil and Magnesium tightens it, in a heavy clay soil you may want 70% or even 80% Calcium and 10% Magnesium; in a loose sandy soil 60% Ca and 20% Mg might be better because it will tighten up the soil and improve water retention. If together they add to 80%, with about 4% Potassium and 1-3% Sodium, that leaves 12-15% of the exchange capacity free for other elements, and an interesting thing happens. 4% or 5% of that CEC will be filled with other bases such as Copper and Zinc, Iron and Manganese, and the remainder will be occupied by exchangeable Hydrogen , H+. The pH of the soil will automatically stabilize at around 6.4 , which is the "perfect soil pH" not only for organic/biological agriculture, but is also the ideal pH of sap in a healthy plant, and the pH of saliva and urine in a healthy human.

So we are looking at two new things so far:

1) The Cation Exchange Capacity, and

2) The proportion of those cations in relation to each other: the percent of base saturation (% base saturation) and their effect on pH.

We are also looking at two old familiar things, clay and soil organic matter, and these last two need a bit more clarification.

How Clay and Humus Form

Clay particles are really tiny. They are so small that they can't even be seen in most microscopes. They are so small that when mixed in water they may take days, weeks, or months to settle out, or they may never settle out and just remain suspended in the water. A particle that remains suspended in water like this, suspended but not dissolved, is known as a colloid. Organic matter, as it breaks down, also forms smaller and smaller particles, until it breaks down as far as it can go and still be organic matter. At that stage it is called humus , and humus is also a colloid; when mixed into water humus will not readily settle out or float to the top. Colloids, because they are so small, have a very large surface area per unit volume or by weight. Some clays, such as montmorillonite and vermiculite, have a surface area as high as 800 square meters per gram, over 200,000 square feet (almost five acres) per ounce! The surface area of fully developed humus is about the same or even higher. Other clays have a much lower surface area; some clays actually have a very low exchange capacity, while humus always has a high exchange capacity.

Mineral soils are formed by the breakdown of rocks, known as the parent material. Heating and cooling, freezing and thawing, wind and water erosion, acid rain (all rain is acid; carbon dioxide in the air forms carbonic acid in the rain), and biological activity all break down the parent material into finer and finer particles. Eventually the particles get so small that some of them re-form, that is they re-crystallize into tiny flat platelets and become colloidal clay, made up mostly of silica and alumina clay particles aggregated into thin, flat sheets that stack together in layers.

Clay "History"

How old a soil is usually determines how much clay it has. The more rainfall a soil gets, the faster it breaks down into clay. Arid regions are mostly sandy and rocky soil, unless they have areas of "fossil" clay. River bottoms in arid regions will often have more clay because the small clay particles wash away easily from areas without vegetation cover. As noted above, clays tend to stick together in microscopic layers. Newly formed clays will often be made up of layers of silica and alumina sandwiched with potassium or iron. On these young clays, the only available exchange sites are on the edges. As the clays age, the "filling" in the sandwich gets taken out by acid rain or soil life or plant roots, opening up more and more negatively charged exchange sites and increasing the exchange capacity. Eventually these clays become tiny layers of silica and alumina separated by a thin film of water. These are the expanding clays; when they get wet they swell, and when they dry out they shrink and crack deeply. Because these expanding clays have exchange sites available between their layers and not just on the edges, they have a much greater exchange capacity than freshly formed clays.

One of the fastest ways to age a clay and reduce the soil's exchange capacity is to use Potassium Chloride fertilizer, KCl. KCl does this by refilling the space between the clay layers with locked-in Potassium and by damaging the edges of the clay layers so that the exchange sites are no longer available. KCl is the cheap Potassium fertilizer used in most commercial mixes; not only does it destroy the exchange capacity of your soil, but the high Chlorine content kills off soil life. It is difficult to have a mineral balanced, biologically active, healthy soil if one is using much Potassium chloride.

In the southern half of the USA, the age of the clay fraction of the soil generally increases going from West to East. The arid regions, from California to western Texas, are largely young soils, containing a lot of sand and gravel and some young clays without a lot of exchange capacity. The central regions, from West-central Texas and above into Oklahoma, Kansas, and Nebraska, contain well-developed clays with high CEC. Moving East, the rainfall increases, the soils are older, and the clays are generally aged and have lost much of their ability to exchange cations. Across Louisiana, Mississippi, Alabama, and Georgia the clays have been rained on and leached out for millions of years. Their reserves of Calcium and Magnesium are often long gone. The northern tier states, from Washington in the West to Pennsylvania and New York in the East were largely covered with glaciers as recently as 10,000 years ago, which brought them a fresh supply of minerals, and clays of high exchange capacity are common.

Organic Matter and Humus

Regarding soil organic matter (SOM) and humus, obviously any area that gets more rainfall tends to grow more vegetation, so the fraction of the soil that is made up of decaying organic matter will usually increase with more rainfall. Breakdown of organic matter is largely dependent on moisture, temperature, and availability of oxygen. As any of these increase, the organic matter will break down faster. Moisture and oxygen being equal, colder northern areas will tend to build up more organic matter in the soil than hotter southern climates, with one extreme being found in the tropics where organic matter breaks down and disappears very quickly, and the other extreme being the vast, deep peat beds and "muck" soils of some North temperate climates. As always, there are exceptions, such as the everglades of Florida, where lack of oxygen combined with stagnant water have formed the largest peat beds in the world; the area around Sacramento California is another example: there were muck (peat) soils 100 feet deep when that river delta was first farmed by European settlers.

Ordinary organic matter from the compost or manure pile, or the remains of last years' crops, doesn't have much exchange capacity until it has been broken down into humus, and from what we know, the formation of humus seems to require the action of soil microorganisms, earthworms, fungi, and insects. When none of them can do anything with organic matter as food anymore, it has become a very small but very complex carbon structure (a colloid) that can hold and release many times its weight in water and plant nutrients. The higher the humus level of the soil, the greater the exchange capacity. One way to increase humus in your soil is by adding organic matter and having healthy soil life to break it down or to add a soil amendment such as lignite (also known as Leonardite), a type of soft coal that contains large amounts of humus and humic acids. If the mineral balance of the soil is optimal, especially with an adequate supply of Sulfur, any fresh organic matter grown in or added to the soil will tend to form stable humus. Without balanced minerals and adequate Sulfur, much of the organic matter will decompose completely and be off-gassed as ammonia and CO_2.

Variable Exchange Capacity

Humus can have an exchange capacity greater than even the highest CEC clays, but it is a variable exchange capacity that correlates with soil pH. In soils with a pH below 6 there will be an excess of H^+ ions in the soil/water solution and many of the negative – exchange sites will be occupied by acidic cations such as Al^{+++} and Fe^{++}. As soil pH increases due to added Ca, Mg, K, and Na, these Al and Fe ions will combine with negatively charged OH^- ions in the soil-water solution, forming insoluble Aluminum and Iron oxides and freeing up the negatively charged sites on the humus to play a role in nutrient exchange. A high-organic-matter soil will have a low "effective" exchange capacity at low pH, because many of the negative exchange sites will be filled with tightly bound Al and Fe. Adding base cations, especially Calcium, will raise the pH and the Calcium++ ions will displace the Al and Fe with "exchangeable" Ca.

OK, let's pull this information together. We have discovered that:

1) Alkaline soil nutrients, largely Calcium, Magnesium, Potassium, and Sodium, are positively charged cations (+) and are held on negatively charged (-) sites on clay and humus.

2) The amount of humus, and the amount and type of clay, determine how much Cation Exchange Capacity a given soil has.

3) We have also discussed the ideal base saturation percentages of these nutrients which according to the work of Professor Albrecht, is approximately:
65% Ca (Calcium)
15% Mg (Magnesium)
4% K (Potassium),
1-3% Na (Sodium)

4) We have talked a little about the effect of those ratios on soil texture and pH and why they are not hard and fast "rules".

The next step is to understand how the plant, and the soil life, gets those nutrients from the exchange sites, the "exchange" part of the story.

Trading + for +

In the same way that acid rain can leach cations from the soil, plants and soil microorganisms more or less "leach" the cation nutrients from their exchange sites. These alkaline nutrients are only held on the surface with a weak, static electrical charge, i.e. they are "adsorbed". They are constantly oscillating and moving a bit, pulled and pushed this way and that by other charged particles (ions) in the soil solution around them. What the plant roots and soil microorganisms do is exude or give off Hydrogen ions, H+ ions, and if these H+ ions are in high enough concentration in the soil solution that some of them surround the nutrient cation and get closer to the negatively (-) charged exchange site than the nutrient cation is, the H+ ions will fill the exchange site, neutralize the (-) charge, and the nutrient cation will be free of its static bond and can then be taken up by the plant or microorganism.

The way this works specifically with plant roots and microbes is that they expire or breathe out carbon dioxide into the soil. This carbon dioxide (CO_2) combines with water in the soil and forms carbonic acid (H_2CO_3); the H+ Hydrogen ions from the carbonic acid are what replaces the cation nutrient on the exchange site.

A Calcium ion that is held to the exchange site has a double-positive charge, written Ca++. When enough H+ ions surround it that some of them get closer to the exchange site than the Ca++ ion is, two H+ ions replace the Ca++ ion and the plant or microbe is free to take the Ca++ up as a nutrient.

How the CEC is measured, and what to do with that information once you have it.

Exchange capacity is measured in milligram equivalents, abbreviated ME or meq. A milligram is of course 1/1000th of a gram, and the milligram being referred to is a milligram of H+ exchangeable Hydrogen. The comparison that is used is 1 milligram of H+ Hydrogen to 100 grams of soil. If all of the exchange sites on that 100 grams of soil could be filled by that 1 milligram of H+, then the soil would have a CEC of 1. One what? One ME, one milligram equivalent (meq), the ability to adsorb and hold one milligram of H+ Hydrogen ions.

Let me repeat that: 100 grams of a soil with a CEC of 1 could have all of its negative (-) exchange sites filled up or neutralized by 1/1000th of a gram of H+ exchangeable Hydrogen. If it had a CEC of 2, it would take 2 milligrams of Hydrogen H+, if its CEC was 120 it would take 120 milligrams of H+ to fill up all of the negative (-) exchange sites on 100 grams of soil.

The "equivalent" part of ME or meq means that other positively (+) charged ions could be substituted for the Hydrogen. If all of the sites were empty in that 100 grams of soil, and that soil had a CEC of 1, 20 milligrams of Calcium (Ca++), or 12 milligrams of Magnesium (Mg++), or 39 milligrams of Potassium (K+) would fill the same exchange sites as 1 milligram of Hydrogen H+.

Why the difference? Why does it take 20 times as much Calcium as Hydrogen, by weight? It's because Calcium has an atomic weight of 40, while Hydrogen, the lightest element, has an atomic weight of 1. One atom of Calcium weighs forty times as much as one atom of Hydrogen. Calcium also has a double positive charge, Ca++, Hydrogen a single charge, H+, so each Ca++ ion can fill two exchange sites . It only takes half as many Calcium ions to fill the (-) sites, but Calcium is 40 times as heavy as Hydrogen, so it takes 20 times as much Calcium by weight to neutralize those (-)

meq/100g vs **cmol$_c$/kg**

The current modern notation for scientific audiences is cmol$_c$/kg (centimoles of charge per kg soil). The "c" subscript before the slash in cmol$_c$/kg denotes "charge". The magnitude of the numbers remains the same. 10meq/100g = 10 cmol$_c$/kg. Many soil testing laboratories still use meq/100g, and we will be using meq/100g in this book because that is the notation used by Albrecht and what will be found in the older research that much of our knowledge of exchange capacity is based on.

charges, or 12 times as much Magnesium, atomic weight 24 (Mg++, also a double charge), or 39 times as much Potassium+. (Potassium's atomic weight is 39, and it has a single positive charge, K+, so it takes 39 times as much K+ as H+ to fill all the exchange sites, once again by weight.) The amount of + charges, the quantitiy of atoms, of K+ or H+, is the same.)

What We Have Learned

We have now learned the basics of CEC, cation exchange, in the soil. 1) Clay and organic matter have negative charges that can hold and release positively charged nutrients. (The cations are adsorbed onto the surface of the clay or humus) That static charge keeps the nutrients from being washed away, and holds them so they are available to plant roots and soil microorganisms

2) The roots and microorganisms get these nutrients by exchanging free hydrogen ions. The free hydrogen H+ fills the (-) site and allows the cation nutrient to be absorbed by the root or microorganism.

3) The unit of measure for this exchange capacity is the milligram equivalent, ME or meq, which stands for 1 milligram (1/1000 of a gram) of exchangeable H+. In a soil with an exchange capacity (CEC) of 1, each 100 grams of soil contain an amount of negative (-) sites equal to the amount of positive (+) ions in 1/1000th of a gram of H+.

Base saturation equivalents for H+, Ca++, Mg++, K+ and Na+:

Per 100 grams of soil,1 meq or ME=
1 milligram H+
20 mg of Calcium Ca++ (atomic weight 40)
12 mg of Magnesium Mg++ (atomic weight 24)
39 mg of Potassium K+ (atomic weight 39)
23 mg of Sodium Na+ (atomic weight 23)

Per Acre, to a depth of 6" to 7", 1 meq or ME=
20 lb Hydrogen H+
400 lb Calcium Ca++
240 lb Magnesium Mg++
780 lb Potassium K+
460 lb Sodium Na+

Per 1000 square feet, 6" to 7" depth, 1 meq or ME=
O.46 lb of Hydrogen H+
9.2 lb of Calcium Ca++
5.5 lb or Magnesium Mg++
17.9 lb of Potassium K+
10.6 lb of Sodium Na+

Per Hectare, to a depth of 15cm to 17cm, 1 meq or ME=
20 kg of Hydrogen H^+
400 kg of Calcium Ca^{++}
240 kg of Magnesium Mg^{++}
780 kg of Potassium K^+
460 kg of Sodium Na^+

To convert hectares to 100 m^2 move the decimal point 2 places to the left: 400 kg/ha = 4.0 kg/ 100m^2

Metric Measurements: Kilograms and Hectares:

The convention used for estimating lbs/Acre in the English/Avoirdupois system is that the top 6" to 7" (15 to 18 cm) of an acre of soil weighs 2 000 000 (two million) pounds, so one part per million (1 ppm) = 2 lbs/acre.

The convention used for estimating kilograms per hectare (kg/ha) is that the top 15 to 18 cm (6" to 7") of a hectare of soil weighs 2 000 000 kg, so 1 ppm = 2 kg/ha.

Considering the huge variance in soil densities, from light weight peat-type soils to heavy clays, unless one wishes to dig up, dry, measure, and weigh a volume sample of the particular soil they are working with, it's safe enough for agricultural purposes to simply say:

1ppm = 2 lb/acre = 2 kg/hectare

1ppm = 20g/1000ft^2 = 20g/100m^2

When calculating soil amendments, be conservative. If you think the amount you are putting on may be too much, use less. It's a lot easier to add more than it is to take something out after adding too much.

To calculate CEC accurately, see the appendix section "Calculating TCEC".

The Ideal Soil Chart (Agricola's Best Guess v 2.0 January 2014)

Based on a Soil Test using the Mehlich 3 method

Organic Matter (OM)	2% — 10%	Depending on climate
pH	6.4 – 6.5	Balance the minerals and pH will take care of itself
Primary Cations as % of Cation Exchange Capacity (CEC) **See appendix "Calculating TCEC" p 125**		
Calcium (Ca)++ min 750ppm	60% — 85% (Ideal 68%)	Ca & Mg together should add to 80% of exchange capacity in most agricultural soils pH 7 and lower
Magnesium (Mg)++ min 100ppm	10% — 20% (Ideal 12%)	
Potassium (K)+ min 100ppm	2% — 5% (Ideal 4%)	See Phosphorus (P)
Sodium (Na)+ min 25ppm	1% — 4% (Ideal 1.5%)	Essential for humans and animals
Hydrogen (H)+	5% — 10% (Ideal 10%)	A lone proton. The "free agent"
Primary Anions		
Phosphorus P- min 100ppm	P = **Ideal** K by weight (ppm) **BUT: phosphate** (P_2O_5) should be **~2X potash** (K_2O)	Needs a highly bio-active soil to keep it available.
Sulfur S - - min 50 ppm	1/2 x **Ideal** K up to 300 ppm	Need for Sulfur amino acids Conserves soil N and Carbon
Secondary elements		
Iron(Fe) + min 50ppm Manganese(Mn) + min 25ppm Zinc (Zn) + min 10ppm Copper (Cu) + min 5ppm	Fe: 1/3 to 1/2 x **Ideal** K Mn: 1/3 to 1/2 x Fe **Zn: 1/10 x P** (up to 50ppm) Cu: 1/2 x Zn (up to 25ppm)	Iron and Manganese are twins/opposites and synergists, as are Copper and Zinc.
Boron $B^{3+ \text{ or } -}$ (cation or anion) min 1ppm	1/1000 of Calcium (max 4 ppm)	Essential for Calcium utilization. Calcium transports sugars
Chlorine (Cl)- min 25ppm	1x to 2x Sodium	Essential, but ages clays rapidly when used in large amounts
Silicon $Si^{4+ \text{ or } -}$ (cation or anion)	Ideal unknown. Si is the most abundant mineral in most soils. Active soil biology and balanced mineral chemistry will ensure availability.	
Micro (trace) Elements		
Chromium Cr- Cobalt (Co)+ Iodine (I)- Molybdenum Mo- Selenium (Se)- Tin (Sn)+ Vanadium (V) + Nickel (Ni) + Fluorine (F) –	All of these are essential in small amounts. 0.5 - 2ppm is enough. Some of the micro elements (e.g. Mo, Se) can be toxic to plants and soil organisms in quantities above 1-2ppm. Use Caution when applying micro/trace elements in purified forms	There are probably 30 or so other elements needed to grow fully nutritious food. Sources are amendments such as seaweed, rock dust, ancient seabed or volcanic deposits, rock phosphate, greensand etc

Plants need at least 17 of the 23 elements listed above, as well as Nitrogen, Carbon, Hydrogen, and Oxygen.

Interlude 1:Introduction to Agricola's Best Guess

Notes on The Ideal Soil Chart version 2.0 January 2014

The Ideal Soil chart evolved from the research of Dr William Albrecht and his associates at the University of Missouri Agricultural Experiment Station in Columbia, Missouri from the 1920s through the early1960s.

William A. Albrecht PhD (1888–1974), Chairman of the Department of Soils at the University of Missouri 1938-1959. President of the Soil Science Society of America 1939.

"He was a very serious man, but very friendly and helpful, especially when it came to helping farmers. He used to say that, when he was young, he had a job cleaning out the offices for a medical doctor of whom he thought a tremendous amount. It was because of that man the he went to college to study medicine. In his pre-med years he took some plant physiology and soil courses, and his interest began to grow. He said that he became disillusioned with medicine when he realized that they were more interested in making money than helping people. He also said "I realised I could help more people through soil science because of the link to health than I could from becoming a medical doctor" Long-time advocate of the Albrecht method Neal Kinsey interviewed in the book "Nutrition Rules" by Graeme Sait, 2003, p13.

"In terms of his standing in the academic world and the farming community itself - and remember , he travelled and addressed congresses throughout the world - I believe that impact at that time was probably greater still than now. The reason I say that was because there were quite a number of influential people that were ready to back that program in the 1940s, '50s, and '60s. It has been said that, had Thomas Dewey won the election instead of Harry Truman, the fellow that would have been Secretary of Agriculture, was a journalist called Louis Bromfield. This guy actually used the Albrecht system and wanted to see it promoted. If Louis Bromfeild had become Secretary of Agriculture, he intended to introduce Albrecht's system as the agricultural fertiliser program for all of the United States. If that had happened, we would have had a much different outcome today." Sait, op cit, p 12.

Q: There are a number of conspiracy theories related to the deposition [deposing] of Albrecht as head of the Department of Soil Science in Missouri. Can you throw any more light on the subject?

A. I'm afraid I can't comment on those theories, but I do know that jealousy and competition within the University may have also played a part. I once sat down and talked with one of Dr Albrecht's closest associates at the University, and he told me that he will always remember a meeting addressed by Dr Albrecht and attended by the Dean of the School of Veterinary Medicine. At that meeting, Dr Albrecht suggested that "*if we could correct our soils, we would correct many of the problems we are having with animal health*." this colleague believed that the Dean became convinced that, if the soils program became really successful, then this would reduce the need or importance of veterinary science. From that day on the Dean was strongly opposed to the soils department, and the veterinary school always had far more money. Charles Walters, who was a very close friend of Albrecht's, tells that, when they actually asked the doctor to step down from the soils department, he was told that "*we need someone who is less of a research scientist and more of a fundraiser*."

In June of 1938, the American Journal of Botany published a bombshell paper entitled *The Colloidal Clay Fraction of Soil as a Cultural Medium* by Wm. A. Albrecht and T. M. McCalla .The paper clearly demonstrated that, contrary to the prevailing wisdom, neither plants nor bacteria were limited to taking in nutrients solely from the soil-water solution; on the contrary they were perfectly capable of obtaining nutrient ions directly from colloidal clay on which the nutrients were adsorbed.

EXPERIMENTAL STUDIES USING COLLOIDAL CLAY MEDIA - Recent studies have shown that the clay fraction of the soil may serve as a source of nutrients. As evidence of this possibility, experiments carried out to date using nutrients adsorbed on the clay and thus not water soluble show that such can be (a) removed by soil microorganisms using them as a source of energy, as the ammonium ion, for example, is oxidized to nitrates; (b) used by soil microorganisms in their regular metabolic processes

for growth, as calcium, for example, is taken by legume bacteria (Rhizobia); and **(c) taken up by plants for their regular growth processes in quantities** related to the supply which is **determined either by the total clay or by the degree of saturation of the clay with the respective nutrient ions**. [emphasis added]
The Colloidal Clay Fraction of Soil as a Cultural Medium by Wm. A. Albrecht and T. M. McCalla, American Journal of Botany, Vol. 25, No. 6 (Jun., 1938), pp. 403-407

For the preceding century the conventional wisdom had been that nutrients must be in the soil-water solution to be available. There had been countless attempts to achieve solutions with a high enough concentration of Calcium, for instance, to grow Rhizobia (the Nitrogen fixing bacteria that form nodules on the roots of legumes) under laboratory conditions, but it had proven impossible. The maximum that could be dissolved in water was 1.3 meq (milligram equivalents) of Calcium per liter.. By the simple method of saturating a colloidal clay with Calcium, Albrecht and McCalla could create a growing medium with 6 meq or more of Ca.

Even though Ashby's medium as commonly used for Rhizobium growth in the laboratory is saturated with calcium carbonate, it contains per liter only 1.3 M.E. of soluble calcium.

Fig. 3. Growth variation with increasing degree of calcium saturation of clay (left to right) and constant amounts of adsorbed calcium. This variation occurs regardless of variation in acidity by hydrogen presence, or of neutrality and of variable barium supply.

The Ideal Soil Chart

Over the past eight years the Ideal Soil mineral ratios have been applied to every imaginable soil type on many hundreds of farms and gardens around the world. It has been proven safe, dependable, and highly effective for growing nutrient dense food crops in all climates. Version 1.0 was assembled in 2005; the first published version was v 1.8 in 2008. The present version 2.0 has only a few minor changes and refinements from v 1.8; it may be used with confidence.

Everything on the chart is related to everything else based on the Cation Exchange Capacity (CEC) of the soil you are working with. Because of that, the starting point is a reliable estimate of the soil's CEC, percent of saturation of that CEC with the base cations Calcium, Magnesium, Potassium and Sodium, and the ratio of those cations to each other.

The first step is to **get a professional laboratory soil test**. One really should have a soil test before adding any minerals at all. You need to know what you are starting with.

The recommended soil test for balancing minerals according to the Ideal Soil method is **the Mehlich 3 test**. The Mehlich 3 or M3 soil test is a strong acid (pH 2.5) extraction. It will measure not only the minerals that are readily available, but also those that are potentially available, the reserves. The Mehlich 3 test is available from most modern soil testing labs.

Two other soil tests are in common use around the world, the **Ammonium Acetate 7.0 pH** test and the **Morgan or Modified Morgan** test.

Although it is useful for measuring readily available nutrients, due to the higher pH (4.8) of the Morgan extracting solution, the **Morgan test results** are not suitable for estimating CEC or balancing the Base Cation Saturation Ratios, nor does the Morgan solution extract sufficient Phosphorus, Iron, Manganese, Copper, or Zinc to give an accurate assessment of the soil reserves.

For soils of pH 7 and below, the **Ammonium Acetate pH 7.0** test will give a good estimate of CEC and the Base Cation Saturation Ratios, but like the Morgan tests, the AA 7.0 pH solution does not extract enough of the secondary minerals to accurately show soil reserves.

For best results, you want a Mehlich 3 soil test with amounts in parts per million ppm for all of these mineral elements:

Primary Cations	Primary Anions	Secondary Elements
Calcium Magnesium Potassium Sodium	Phosphorus Sulfur	Boron Iron Manganese Copper Zinc

These are minerals whose function we understand well and it is essential that they all be in your soil in sufficient quantities. You do not need to know the amounts of the micro trace elements to start with (those at the very bottom of Agricola's chart), and ordinary soil tests don't measure them anyway.

The laboratory may also estimate the **CEC** (cation exchange capacity) and the **base saturation percent** of Calcium, Magnesium, Potassium, Sodium and Hydrogen in your soil sample. Unless you know that the method the lab is using to calculate CEC is the same one described in this book in the appendix on **Calculating TCEC**, it would be best to calculate it yourself based on the amounts of Ca, Mg, K, and Na and soil pH at 1:1 soil:water by weight.

If the soil pH is above 7.0, the soil probably contains free, undissolved Calcium and/or Magnesium carbonates, and both the AA 7.0 test and the Mehlich 3 test are likely to dissolve them and give too high a reading for Ca and Mg, leading to an overestimate of CEC. For soils above pH 7.0, the Ammonium Acetate pH8.2 test (AA8.2) is recommended for determining CEC and base saturation percentages only. You will still need the Mehlich 3 test results for all other mineral elements. See Chapter 9 on Calcareous and High pH soils for more detail.

In high doses, many mineral elements can be toxic to people, animals, plants and soil organisms. This is true regardless of whether they are in a naturally occurring or purified, concentrated form. Keep them out of ponds and streams. *Any* mineral, if used in excess, can throw things out of balance, so take it easy. It is much easier to put them in than to get them back out of the soil.

Caution When Applying Mineral Amendments

High levels of some minerals in the soil may inhibit sprouting of seeds. Boron is definitely known to do this. High levels of free minerals (not biologically assimilated) can also "plug up" the vascular systems of young plants, stunting their growth. Seeds may sprout fine but stall out after the first set of true leaves. This seems to be particularly true after adding high amounts of Calcium. For these reasons it is best to wait until the minerals are chemically and biologically a part of the soil before starting seeds in it. Transplants usually do fine if you wait a week or so after adding large quantities of minerals before replanting them, and we have seen no problems with established plantings, trees, or pastures. Adding minerals in the fall or in the early spring works best.

If minerals are added directly to potting mixes the mix should be moistened after mixing in the minerals, and it is best to give it a little time, a week or so, to "settle in" before the potting mix is used. Adding a biological activator such as beneficial bacteria or fungi to the mix can greatly speed up the process.

The mineral concentrations shown on this chart are perfectly safe for plants once they are assimilated into the living soil. If the chart's guidelines are followed you won't end up with too much of anything–many soils naturally contain higher levels of available minerals than the chart calls for.

If you are a cautious or doubtful person, or the expense seems too great, you may choose not to balance the minerals on the whole farm or the whole garden or pasture at once. Start with maybe one-half of the area and see how things go, or divide it into two or more parts and treat them slightly differently, for instance putting the whole amount called for on one part and only half that amount on the other. It is always a good idea to leave a small representative area untouched as a control. After a year or so you will enjoy pointing out that area and saying "See, that's what I started with!"

Don't expect immediate and fantastic results from adding some minerals to your soil. It takes time for them to work their way into the living systems of the soil. As the minerals settle into the soil ecology, some will become available to the plants and soil microorganisms and others may get tied up for a while. Adding a little bit of a badly needed mineral nutrient to the soil may greatly increase microorganism and fungal activity, and may catalyze the release of other previously bound-up minerals.

If you are serious about gardening or farming and having the healthiest soil and plants possible you will want to get a soil test at least once a year. Twice a year, in the spring and in the fall is even better. The spring test will show you what you should apply for this year's crop, and the fall test will tell you what to add to settle in over the winter. Calcium and Magnesium, for example, become much more bio-available if they are spread on top of the soil in the fall and allowed to leach into the soil with the winter's rain or snow.

When the mineral balance of the soil is brought into line with the Ideal Soil Chart, the pH will self-correct to what is perfect for your soil and climate.

Mother Nature and the soil are very forgiving and you do not have to be *exact* in these proportions. It would be unlikely to find two soil samples taken one foot apart that were identical. The soil test will give you the general idea, and as long as you go slow and take it easy everything will be fine.

If a large area is to be balanced and cost precludes applying the full amounts of all of the needed minerals, start with the most important cation minerals, Calcium and Magnesium. They are fully as important in the soil as they are in the human body and the least expensive to buy. In a very loose and sandy soil with a low exchange capacity you will want about 60% Ca saturation and 20% Mg saturation, in a heavy clay soil with a high exchange capacity, 70% to 80% Ca to 10% Mg. This is because the higher the ratio of Calcium to Magnesium, the looser the soil gets, and as the Magnesium portion gets higher, the soil gets tighter. A higher level of Mg will pull a loose sandy soil together; a higher level of Ca will open up a dense, heavy soil.

Calcium sources: Agricultural sweet lime (Calcium carbonate) and gypsum (Calcium sulfate) are the preferred sources of calcium. Gypsum supplies readily available Calcium, and is also a good source of Sulfur, an element that is seriously lacking in most agricultural soils. Agricultural lime supplies Carbon as well as Calcium. Carbon helps make a soil less sticky. If you already have plenty of Carbon in your soil as organic matter, but are low on Sulfur, gypsum is a better bet. The various rock phosphates and regular superphosphate also contain significant Calcium, but their Calcium content is chemically bound to Phosphorus and is not available in exchangeable form, so should not be considered as part of the Calcium being added to balance the CEC ratios.

As a rule, don't use **Dolomite** lime, regardless of what you may have read in various gardening books, unless you are sure that you need Magnesium. Dolomite is a high Magnesium limestone. Using dolomite will tighten the soil, reducing air in the soil and inducing anaerobic alcohol fermentation or even formaldehyde preservation of organic matter rather than aerobic decomposition. If the soil test calls for more Magnesium, Magnesium sulfate (Epsom salts) or K-Mag (also known as Sul-Po-Mag, sulfate of potash magnesia, or Langbeinite), are generally safer and quicker acting sources of Magnesium than dolomite. Magnesium oxide is the purest and quickest acting Magnesium additive, but is not presently allowed under USDA NOP organic rules, for some reason. About the only time dolomite lime might be called for would be if the soil already had too high a level of Sulfur to use Magnesium sulfate (Epsom salts) or K-Mag, or if other sources of Magnesium were not available. If one is not concerned with being "certified" organic under USDA rules, Magnesium oxide is the best bet. MgO (Magnesium Oxide) is around 50% Mg, a much higher percentage than dolomite lime (13% Mg) or Epsom salts (10% Mg) so it is also a much cheaper source of Mg. If you are not concerned about being certified by the government, I would recommend using MgO.

Agricola's chart says that Phosphorus and Potassium should be equal, but that's not as simple as it looks. On a bag of fertilizer sold in the USA or Canada, such as 10-10-10, the numbers stand for NPK, in that order. The N number is for Nitrogen, but the P number actually stands for *phosphate,* P2O5, and the K number stands for *potash,* K2O (K is from the German word Kalium, meaning Potassium). Phosphate is 44% Phosphorus, while potash is 83% Potassium. So, one needs about twice as much phosphate as potash for the P and K to be equal. A 10-20-10 or a 2-4-2 fertilizer would have that correct ratio.

It is also important to know what form of P and K the soil test is listing. Some labs give the P number as phosphate, P2O5, so you can take that times .44 and find the actual amount of Phosphorus in your soil. Some give the K number as K2O, some as actual K. It is best to request that the lab results be listed in parts per million of the pure element being measured.

.

Although this chart emphasizes *minerals,* you would not have much luck trying to grow food in a soil that wasn't bio-active even if it contained the perfect mineral balance. The goal is to get these minerals into the soil in a biological or at least bio-available form. We add them and let the soil life assimilate them over time.

In some cases minerals may be added to the compost pile to start the bio-availability process, but it's a good idea to keep *good* records of how much of *what* is in *which* pile. For example, one could mix a bag of rock phosphate into a good sized compost pile , but it would be nice to know just how much Phosphorus, Calcium etc was in the bag to start with, and that it was all in *that* pile and could be spread over X amount of area.

Know *all* the ingredients of anything you add to the soil if at all possible. How much Cadmium (a toxic heavy metal) does that phosphate rock have in it?

Glacial rock dust, granite dust etc. cfan be great sources of fresh minerals, but they can't be relied on to supply the primary cations and anions. Most of them have low enough numbers of the major nutrients that they won't throw things out of balance, though, and because they are freshly ground up and sharp grains of rock, they will increase the energy level in the soil. Both heat and electrical charge concentrate at sharp points.

Handy Facts:
The top 6 to 7 inches of an acre of soil is assumed by convention to weigh two million pounds (2 000 000 lbs). The top 15 to 17 centimeters of one hectare of soil is assumed to weigh two million kilograms (2 000 000 kg).This is referred to as the *plow layer*, and is where most of the growth happens and where most of the available nutrients are.

One part per million (1 ppm) of the plow layer equals two (2) pounds per acre or two (2) kilograms per hectare:
1ppm = 2 lbs/acre
1ppm = 2 kg/ha
An acre is 43 560 square feet, or close to 45 000 ft^2
A pound is 453 grams, or about 450 grams.

One part per million = approx. 2 grams per 100 square feet.
1 pound/acre = 1 gram/100 ft^2
1ppm = 2 grams/100 ft^2
1ppm = 2 grams per 10 $meters^2$

(To calculate Total Cation Exchange Capacity TCEC accurately, see the Appendix section on Calculating CEC)

A printable periodic table of the elements in .PDF form may be found here:
http://www.webelements.com/nexus/sites/default/files/webelements_table_5sf_2012-06-07.pdf

Interlude 2: Conventions Used in This Book

The Decimal Mark (Decimal Point)

The decimal mark is what separates whole units from tenths, hundredths etc. In European countries (except the UK and Ireland), in South America (except Ecuador), and in South Africa, the decimal mark is a comma: **1,01** means one and one one-hundredth (1+ 1/100th) . In North America, India, East Asia, and Australia the period or dot is used instead: **1.01** means one and one one-hundredth. This gets even more confusing when commas or dots are used to separate large numbers into groups. **1.000.000,00** means one million and no hundredths in many countries; the same would be written **1,000,000.00** in English speaking and many other countries.

The 22nd General Conference on Weights and Measures declared in 2003 that "the symbol for the decimal marker shall be either the point on the line **(.)** or the comma on the line **(,).** It further ruled that "numbers may be divided in groups of three in order to facilitate reading; neither dots nor commas are ever inserted in the spaces between groups". Thus one million may be written **1 000 000,00** or **1 000 000.00**.

In this book we will use the dot or period (**.**) to separate the whole numbers from the fractional decimals, and will separate groups of three with only a space, e.g.:

1 000 000.001 (one million and one one-thousandth)

Weight of Soil Per Unit of Area

In the examples in this book we will generally be dealing with the "plow layer", the upper horizon, the topsoil. We will assume that the **plow layer** of a **hectare** of land **weighs 2 000 000 (two million) kg**, and the **plow layer** of an **acre** of land **weighs 2 000 000 lbs.** In most soils this will be **the top 15 to 17 cm of one hectare**, or **the top 6 to 7 inches of one acre**.

Parts Per Million ppm

In the next several chapters we will be writing a soil mineral Rx based on the results of a Mehlich 3 soil test. The units we will be using are **parts per million (ppm)** of **weight**.

One gram is one ppm of one million grams (1000 kg or 1 metric ton)
1 milligram is 1 ppm of 1 kilogram
1 pound is 1 ppm of 1 000 000 pounds

On a large field one will be calculating for 2 million kg of soil per hectare (2 million lbs/ac) or more; for a greenhouse one may be calculating how much to add to a few kilograms of potting soil mix. Working with ppm allows us to do the calculations once, in a simple manner, and then apply them to any units of weight we wish: grams, pounds, ounces, kilograms, or tons, and to any area: ares, hectares, square meters, acres, square feet, cubic meters or cubic yards.

To readily use the calculations in the following chapters, **if your soil test results are not already in ppm, you will need to convert the numbers you have to ppm.**

Many soil testing laboratories already report their test results in ppm; some, mostly in the USA, report in pounds per acre. If the lab you are using usually reports in lbs/acre (or kg/hectare) you can ask them to print your soil report in ppm, or do the conversion yourself, based on the convention **that for a depth of 6" or 15cm, 1ppm = 2 kg/hectare (or 2 lbs/acre)**. Just divide the lbs/acre or kg/hectare result by 2.

1000 lbs/acre ÷ 2 = 500 ppm
1000 kg/hectare ÷ 2 = 500 ppm

If the soil sample was taken to a depth other than 6" or 15cm, and the lab results are in lbs/acre or kg/hectare, you will need to compensate for the difference in weight. If the sampling depth is doubled to 12" or 30 cm, the weight would be 2 x as much,
4 million lbs/acre or kg/hectare, and 1 ppm would be 4 lbs/acre or 4 kg/hectare.

Sample Depth	Weight in kg/ha or lbs/acre	1 ppm = kg/ha or lbs/acre
4" or 10 cm	1 333 333	1.33
6" or 15 cm	2 000 000	2.0
8" or 20 cm	2 666 666	2.6
12" or 30 cm	4 000 000	4.0

Converting Phosphate and Potash into Phosphorus and Potassium

Again mostly in the USA, some labs report Phosphorus as P_2O_5, phosphate, and Potassium as K_2O, potash. If the lab you are using does this, the conversion to elemental P and K is:

P_2O_5 (phosphate) x 0.44 = elemental P (Phosphorus)
K_2O (potash) x 0.83 = elemental K (Potassium)

In any event, you will want to end up with all of the lab test data on your worksheet in ppm of the elements tested.

Acres and Hectares

One hectare (abbreviated ha) is 10 000 meters^2 (100m x 100m). 1/100th of that is 100 m^2, an area of measure also known as an Are (pronounced "air"). 100 Ares = 1 hectare. This makes the conversion from kg/hectare to kg/Are simple: just move the decimal point 2 spaces to the left. 120 kg/ha = 1.20 kg/Are. (1 Are = 100 m^2 = 1076 sq ft)

The **acre** originated in Europe as **the area one man could plow with a team of oxen in a day**. This was measured in chains (a chain is 22 yards or 66 ft long). An acre was (is) defined as an area 1 chain wide and 10 chains long, 22 yards x 220 yards.

1 acre = 66 ft x 660 ft. = 43 560 ft^2.

An acre (abbreviated ac) is about 40% of a hectare – slightly smaller than an American football field, which is 50 yards x 100 yards (45 000 ft^2). The countries that presently use the acre include the United States, Australia, India, Pakistan,

Burma and the United Kingdom (as of 2010, the acre is no longer officially used in the United Kingdom, though is still used in real estate descriptions). It is also still used, to a large extent, in Canada.

An average home vegetable garden is around 25 ft x 40 ft = 1000 ft^2. There are **43.56 1000 ft^2 sections in an acre. To convert lbs/acre into lbs/1000 ft^2, divide lbs/acre by 43.56.**

Web Links in Hardcopy and Ebook
Letters that are grey-colored or **underlined** in the hardcopy version of The Ideal Soil are URL **links** to websites or to other pages in The Ideal Soil in the **ebook version**.

meq/100g vs **$cmol_c$/kg**
The accepted modern notation for scientific audiences is $cmol_c$/kg (centimoles of charge per kg soil). The “c” subscript before the slash in $cmol_c$/kg denotes “charge”. The magnitude of the numbers remains the same. 10meq/100g = 10 $cmol_c$/kg. Many soil testing laboratories still use meq/100g, and we will be using meq/100g in this book because that is the notation used by Albrecht and what will be found in the older research that much of our knowledge of exchange capacity is based on.

Capitalization of Names of Elements
The names of the elements are capitalized in this book. The various chemistry terms that refer to their combination with other elements are not capitalized; e.g. Sulfur in Sulfur trioxide SO_3 will be capitalized; sulfate in Magnesium sulfate $MgSO_4$ will not be capitalized because sulfate refers to the SO_4 molecule, not the pure element S.

Primary, Secondary, and Micro Elements
The common English agronomic terms Major, Minor, and Trace elements can be confusing because they don’t accurately describe either the importance or the relative amounts of the element in the soil or the plant. Neither do they translate easily into some languages or retain the same meaning when translated as they do in English.

Another common usage is Macro- Micro- and Trace elements, which is not much better than Major and Minor. Using the prefixes Macro- Meso- (middle, medium) and Micro was considered but rejected because it would entail introducing “meso”, a term not in common usage in many languages.
In the event, this book will use the terms **Primary**, **Secondary**, and **Micro** to describe and rank mineral elements in a general way by their required amounts in the soil and percentage occurrence in the crops grown. **Primary**, **Secondary**, and **Micro** are not meant to reflect the relative importance of an element in the health of living things, but only their approximate ratio of abundance in a fertile, balanced soil.

Essential Primary, Secondary, and Micro Elements

Primary Elements Needed in large amounts in soil	**Secondary Elements** Needed in lesser amounts in soil	**Micro Elements** Needed in very small amounts in soil
Anions Sulfur S Phosphorus P Nitrate NO_3^-* **Cations** Calcium Ca Magnesium Mg Potassium K Ammonium NH_{4+}* (*A molecule, not an element)	**Anions** Boron B Chlorine Cl Silicon Si **Cations** Sodium Na Iron Fe Manganese Mn Copper Cu Zinc Zn	Cobalt Co Selenium Se Molybdenum Mo Nickel Ni Vanadium V Chromium Cr Yttrium Yt Cesium Cs Strontium Sr Fluorine Fl Iodine I Titanium Ti Lanthanum La Cerium Ce ....and many more

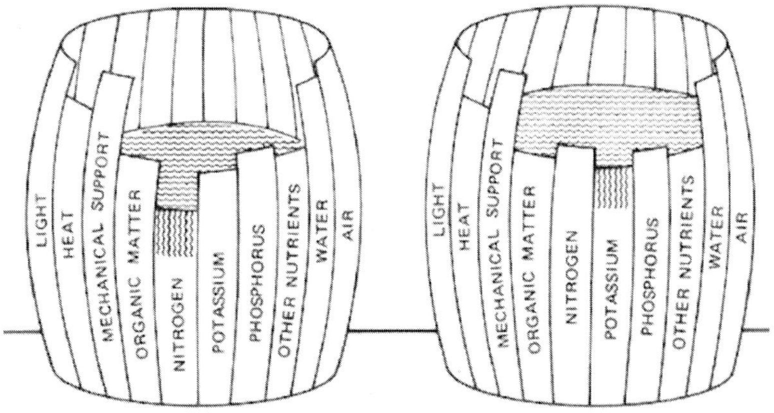

Figure 1: Liebig's Law of the Minimum showing most of the factors that can limit the growth of crops.

To the left is an illustration of Justus von Liebig's "Law of the Minimum" showing most of the factors that limit growth, from light to heat to soil minerals and air. Nitrogen is the short stave, the limiting factor, in the barrel on the left. The barrel on the right shows Nitrogen brought up to the desired level; now Potassium is the limiting factor. It can truly be said that all of the essential elements are equally important. Whichever one is in shortest supply will limit the health and growth of the entire organism.

A printable periodic table of the elements in .PDF form may be found here:
http://www.webelements.com/nexus/sites/default/files/webelements_table_5sf_2012-06-07.pdf

Chapter 3

Balancing the Cation Nutrients (See appendix "Calculating TCEC")

Calcium Ca++, Magnesium Mg++, Potassium K+ and Sodium Na+

Please read the previous section "Conventions Used in This Book", especially the part about **parts per million**. We will be working with ppm in the next 4 chapters.

The soil is the storehouse of fertility, or at least it should be. The minerals that the plants need in order to grow and reproduce another healthy generation all come from the soil and are stored there, in or on one of the following forms:

- The clay fraction
- The organic fraction: both living and dead/decaying
- Rock minerals of various sizes and types, from silt to boulders
- The soil/water solution: dissolved nutrients, easily available, also easily leached out.

The storehouse capacity we are concerned with right now is a property of the first two items on the list, the **clay** and **organic** fractions. The soil/water solution and the rocks are not considered part of the exchange capacity.

What we wish to do is to load the storehouse up with the mineral nutrients in the proper balance. This will accomplish a number of things:

- On a physical level, it will keep the soil loose and friable so air, water, plant roots and soil organisms can move through it freely
- On the level of chemistry, it will allow the acid/alkaline balance, the pH, to self adjust, ideally to around pH 6.4. A pH of 6.4 is where the maximum amount of nutrients are available.
- By having the proper balance of cation nutrients filling the exchange sites, those nutrients will be readily available to the plant roots and soil organisms, so they can "trade" Hydrogen H+ ions for the nutrients they need.
- When the cation nutrients are held to the soil colloids (clay and humus) they are not subject to leaching or washing away due to rainfall or irrigation.

Here is the laboratory Soil Test Report we will be working with in this book

Element		Results	Comments
Total Cation Exchange Capacity TCEC		11.4	
pH of Soil Sample		5.58	
Organic Matter %		5.6%	
Anions			
Sulfur S (parts per million ppm)		20	
Phosphorus P ppm		100	
Cations			
Calcium Ca++ ppm	Desired Found Deficit	1250	
Ca Base Saturation 60-70 %			
Magnesium Mg++ ppm	Desired Found Deficit	116	
Mg Base Saturation 10-20 %			
Potassium K+ ppm	Desired Found Deficit	89	
K Base Saturation 2-5 %			
Sodium Na+ ppm	Desired Found Deficit	26	
Na Base Saturation 1-5 %			
Other Bases		6.2%	
H+ Exch Hydrogen 10-15%		27.0%	
Other Elements ppm			
Boron		0.21	
Iron Fe		50	
Manganese Mn		11	
Copper Cu		1.07	
Zinc Zn		16.4	
Aluminum		1841	Normal

Our overall goal is to feed the soil and the soil organisms that in turn feed the plants. Our specific goal is that the soil should contain perfect nutrition for the crop we wish to grow. In the case of plants grown for food, we also want the crop to contain all of the nutrients essential for the health of the people or animals that will be eating the food being grown. Luckily, most food crops do best with the same soil mineral balance.

The Calcium : Magnesium ratio sets the stage for all of the rest of the elements. If the Calcium level is too high in relation to Magnesium, the soil will be loose but will lose its texture and cohesiveness and water may drain through too easily and be lost. It will also be more prone to erosion from wind or water. If the Magnesium level is too high, the soil will be tight, preventing water and air from moving through easily.

From the Ideal Soil chart, here is the preferred range of base saturation for most plants:

Calcium (Ca)++ min 750ppm	60% — 85% Optimum 68%	Ca & Mg together should add to 80% of exchange capacity in most agricultural soils pH 7 and lower
Magnesium (Mg)++ min 100ppm	10% — 20% Optimum 12%	
Potassium (K)+ min 100ppm	2% — 5% Optimum 4%	See Phosphorus (P)
Sodium (Na)+ min 25ppm	1% — 4% Optimum 1.5%	Essential for humans and animals
Hydrogen (H)+	5% — 10% Optimum 10%	A lone proton. The "free agent"

A very heavy clay soil needs to be loosened up, so one would wish to see a Ca:Mg saturation ratio of perhaps 75% (or even more) Calcium to 10% Magnesium. A very loose sandy soil needs to be tightened up to hold water and prevent erosion; in that case 60% Calcium and 20% Magnesium would be desired.

Heavy clay: 75% (or more) Calcium, 10% Magnesium

Loose sand: 60% Calcium, 20% Magnesium.

At no time do we want the Calcium saturation to be below 60% or the Magnesium saturation below 10% unless we are growing specialty crops such as blueberries or rhododendrons that like a high-Magnesium and somewhat acid soil or certain plants that prefer a very high Calcium "chalky" soil.

In an "ideal" soil that has a good mix of sand, silt, and clay as well as a good level of organic matter, Professor Albrecht determined that the best ratio was 65% Calcium to 15% Magnesium. Further experience has convinced other agronomists that slightly different ratios work better for them. At soilminerals.com we usually recommend a "perfect" ratio of 68% Calcium to 12% Magnesium for soils below pH 7. This seems to be the ideal proportion not only to give the plants and soil life the nutrients they need in the proper ratio, but to keep the soil loose and friable while retaining soil moisture.

Here is our ideal soil cation saturation ratio:

Calcium 68%

Magnesium 12%

Potassium 4%

Sodium 1.5%

Other bases 4% to 5%

Free Hydrogen H+ 10%

A soil with this ratio of cation minerals will self-adjust to a pH of about 6.4 given adequate soil moisture. Most of these elements are being held on the clay fraction in a clay soil, and on the organic (humus) fraction in a sandy or high-organic matter soil such as peat or muck soils. Sand and gravel have almost no exchange capacity. Aged clays, especially in the tropics and subtropics, also have very low exchange capacity. Methods of increasing the exchange capacity of low-CEC soils include adding organic matter, charcoal (biochar), high CEC clay (e.g. montmorillonite/bentonite), or humic acid sources. Biochar and montmorillonite clay will both confer permanent exchange capacity.

~~

Let's take another look at the list from the end of the *Cation Exchange Capacity Simplified* chapter:

meq/100g vs **$cmol_c$/kg**

The accepted modern notation for scientific audiences is $cmol_c$/kg (centimoles of charge per kg soil). The "c" subscript before the slash in $cmol_c$/kg denotes "charge". The magnitude of the numbers remains the same. 1meq/100g = 1 $cmol_c$/kg.

Many soil testing laboratories still use meq/100g, and we will be using meq/100g in this book because that is the notation used by Albrecht and what will be found in the older research that much of our knowledge of exchange capacity is based on.

Per 100 grams of soil,1 milligram equivalent (meq or ME)=
1 milligram Hydrogen H+ or
20 mg of Calcium Ca++ or
12 mg of Magnesium Mg++ or
39 mg of Potassium K+ or
23 mg of Sodium Na+

Again, if you take 100 grams of oven-dry soil, with a CEC of 1, one milligram of free Hydrogen H+ will fill all of the negative exchange sites. As we will be working with parts per million, and 100 grams is 1/10th of 1 kilogram, we multiply the above numbers x 10 to get mg/kg or ppm:

1 meq (or $cmol_c/kg$) =
10 ppm Hydrogen $H+$
200 ppm Calcium $Ca++$
120 ppm Magnesium $Mg++$
390 ppm Potassium $K+$
230 ppm Sodium $Na+$

To convert these numbers to kg/ha or lbs/acre, multiply by 2:
200 ppm = 400 kg/hectare or 400 lbs/acre

So how do we put this information to use?

Let's say we have a soil with a CEC of 1meq. This is a normal soil, not sandy or high in clay. We decide that we wish to end up with 65% of the exchange sites filled with Calcium and 15% with Magnesium, Albrecht's optimum ratio. We know from above that 200 ppm of Calcium will fill 100% of the exchange sites. We want only 65% filled with Ca, so

200 x 0.65 = 130 ppm Ca

We want 15% base saturation of Magnesium, and know from above that 120 ppm Mg will saturate 100%

120 x 0.15 = 18 ppm Mg

130 ppm of Ca and 18 ppm of Mg will give us the 65:15 ratio, on a soil with a CEC of 1.

How about the "ideal" ratio mentioned above, 68% Ca, 12% Mg, 4% K, and 1.5% Na?

Still working with a soil where CEC = 1:

Calcium: 200 ppm x 0.68 = 136 ppm

Magnesium: 120 ppm x 0.12 = 14.4 ppm

Potassium: 390 ppm x 0.04 = 15.6 ppm

Sodium: 230 ppm x 0.015 = 3.45 ppm

Note that Potassium and Magnesium are almost the same weight in parts per million, despite the fact that Magnesium is filling 12% of the EC while Potassium only fills 4% of the exchange sites. With few exceptions, exchangeable Potassium and Magnesium should be approximately equal in weight in the soil.

****

Enough arithmetic for a bit. Let's review why we are doing what we are doing. The primary cations Ca, Mg, K, and Na that we are balancing are all nutrients for plants and animals. They are also chemical elements with their own properties. They interact with each other, and compete for the available exchange sites. Too much of one may mean not enough of another. If the whole CEC were saturated with Calcium, where would the Magnesium be stored? Answer: It wouldn't be stored; it would be either in the soil solution or heading for the water table on its way to the ocean.

As we covered in Chapter 2: Cation Exchange Simplified, these nutrient minerals are held on the clay or humus by a static electric charge. They are positively charged **+**, the sites where they are attracted and held are negatively charged **-**. Plant roots and microorganisms can donate a couple of H+ Hydrogen ions to fill the two negative sites occupied by the Ca++ ion, thus freeing the Ca to be absorbed as a nutrient. That's the exchange: 2H+ for 1Ca++, or 2H+ for 1Mg++.

Or 1H+ for 1K+
or Mg++ for Ca++
or 2K+ for 1Ca++

A cation nutrient that is held to an exchange site, say Ca++, may be exchanged for 2 H+ ions. Over time, as more and more Ca++ ions are exchanged for H+ ions released from plant roots, from soil microorganisms, or simply from free H+ in rainfall, more sites become filled with H+ and the soil becomes more sour or acid. The pH gets lower. A pH of 4.9 would be 44% saturated with Hydrogen, far too acidic for most plants. A pH of 7.0, neutral, would have no exchangeable H+ adsorbed on the colloids at all, and a pH of more than 7.0 would have more **+** minerals available in the soil than the exchange sites could hold.

In a garden, field, or orchard where crops are grown, harvested, and taken away, the mineral nutrients are taken away along with the crop. The same is true for pastures where animals are grazed for milk or meat. In order to continue to raise high-quality nutrient dense food crops, we must replace what has been taken away; we also want to continue to have the proper balance of mineral nutrients available.

Getting Started:
Below are the primary cation results from the soil report we will be using throughout this book. We will use them and the information discussed above to determine the percent of base saturation of the cation nutrients, determine if we

need to add more, and how much we would need to add to achieve the Ideal Soil ratio or any other ratio we want. **The complete soil report is at the end of this chapter.**

Exchange Capacity	Calcium ppm found	Magnesium ppm found	Potassium ppm found	Sodium ppm found
11.4	1250	116	89	26

First we will determine the percentage of base saturation of each of these elements, and then figure out if we need to add more and if so, how much.

The soil's exchange capacity is 11.4 meq. Starting with Calcium, we multiply each element by the amount needed (in ppm) to saturate that 11.4 CEC 100%:

Ca; 11.4 x 200 = 2280 ppm

Mg: 11.4 x 120 = 1368 ppm

K: 11.4 x 390 = 4446 ppm

Na: 11.4 x 230 = 2622 ppm

The table above shows how many parts per million the soil test found. To find out the percent base saturation for each element in our working example, we divide the amount measured by the lab test by the amount needed to saturate 100%:

Ca: 1250 / 2280 = 0.548 or 55%

Mg: 116 / 1368 = 0.0848 or 8.5%

K: 89 / 4446 = 0.020 or 2%

Na: 26 / 2622 = 0.0099 or 1.0%

Percent Base Saturation of Worksheet Sample

Calcium	Magnesium	Potassium	Sodium
55.0%	8.5%	2.0%	1.0%

Recall we want 68% Ca, 12% Mg, 4% K, and 1.5% Na. To calculate what that "ideal" ratio would be in this soil we simply multiply the exchange capacity (11.4meq) by the amount needed to saturate 100% of 1meq, and then multiply that result by the percentage of saturation desired.

68% Calcium: 11.4 x 200 x 0.68 = 1550 ppm

12% Magnesium: 11.4 x 120 x 0.12 = 164 ppm

4% Potassium: 11.4 x 390 x 0.04 = 178 ppm

1.5% Sodium: 11.4 x 230 x 0.015 = 39 ppm

This works the same way no matter what cation you are working with or what percentage of base saturation you desire. If you wanted 4% Na, it would be 11.4 x 230 x 0.04 = 105 ppm and so on.

Next we subtract the amount measured on the soil report from the ideal amount calculated above to find the amount we need to add:

Ca: 1550 - 1250 = 300 ppm

Mg: 164 - 116 = 48 ppm

K: 178 – 89 = 89 ppm

Na: 39 - 26 = 13 ppm

Amount Needed to Raise Base Saturation to Ideal Level

Calcium	Magnesium	Potassium	Sodium
300 ppm	48 ppm	89 ppm	13 ppm

Adding these amounts to the initial values reported from the soil test will give us these totals

Total of Major Cations after Balancing

Calcium ppm	Magnesium ppm	Potassium ppm	Sodium ppm
1250 + 300 =**1550** = 68%	116 + 48 = **164** = 12%	89 + 89 = **178** = 4%	26 + 13 = **39** = 1.5%

In the next chapter we will use the amount of Potassium shown in the table above to determine what our ideal level of Phosphorus should be.

In this chapter we have learned how to use simple arithmetic to calculate percentage of base saturation for the major nutrient cations, and how to figure out what we need to add in order to achieve the percentage we desire. As noted earlier, different soils may need a different ratio than the "ideal" ratio. Heavy clay soils may need up to 80% or more Ca saturation, light sandy soils up to 20% Mg. As a general rule, the amount of K, Potassium, should equal the amount of Mg, Magnesium, *by weight*, but the K level should not be more than 5% of base saturation. Sodium % is not critical, as long as it is above 0.5% and below 5%, however if the % of Na is below 2% it gives us the opportunity to add sea salt to the soil, an excellent source for all of the trace minerals dissolved in the ocean.

Many laboratory soil reports will list % saturation of the cations Ca, Mg, K, and sometimes Na. Unless you already know and trust what method the lab is using for calculating CEC and percent saturation, it is a good idea to double check by recalculating them yourself.

What if the soil test shows levels of the major cations that are already too high?

There are many soils found around the world that have an "inverted" Ca/Mg ratio, with Magnesium higher than Calcium or at least higher than 20% Mg, or Sodium levels above 5%. There are also many soils that naturally have very high Calcium levels (above 80% base saturation) but lack Magnesium. The usual approach to correct this problem is to add elemental Sulfur in one form or another, or a sulfate form of the needed cation. Most soils worldwide are Sulfur deficient, and plants love Sulfur. Depending on the situation, the answer may be to add pure 90% agricultural Sulfur, Magnesium sulfate, Calcium sulfate, Potassium sulfate, or K-Mag (sulfate of potash magnesia). Which one is used will depend upon the overall balance of the major cations.

Calcium too high, Magnesium low: Add Magnesium sulfate, (Epsom salts).

Magnesium too high, Calcium low: Add Calcium sulfate, (gypsum).

Calcium and/or Magnesium high, Potassium low: Add Potassium sulfate

Ca, Mg, and K all high or adequate, high pH: Add 90% agricultural Sulfur

Sodium high: Add the sulfate salt of whichever of the other major cations is deficient, or 90% ag Sulfur.

What adding Sulfur or sulfates does is induce the sulfur to chemically bond with whatever other cation is in excess. For instance, if one adds Magnesium sulfate (Epsom salts), to a soil that is high in Calcium but low in Magnesium, the Sulfur in the Epsom salts will tend to attach to the Calcium, pulling the excess Calcium from the exchange site and leaving an atom of Magnesium. The result is free Calcium sulfate (gypsum) in the soil/water solution. Gypsum is water-soluble and mobile; rain and irrigation water will tend to leach it downwards out of the root zone. Calcium, incidentally, always ends up leaching out of topsoil into subsoil. If one digs a deep trench and tests the soil at different levels in the trench, the highest Calcium levels will be found lower in the trench. That depth depends on the amount of rainfall. In a maritime climate such as the Pacific Northwest USA, where rainfall averages around 100 cm (40 inches) per year, a high Calcium layer will often be found at a depth of around 2.5 m (8 feet). This is much too deep for the roots of most crops to reach. In highly leached soils such as this there is often a Calcium deficiency in the topsoil; that tends to attract extremely deep-rooted plants such as dandelion and Canada thistle. These plants have a long taproot that reaches down to the Calcium layer and pulls the Calcium back up to the topsoil. If you have a dandelion problem in your lawn or garden, suspect Calcium deficiency. In more arid climates such as the American Southwest, the depth of the Calcium layer will be much less, usually 30 to 90 cm (1-3 feet) deep. Because there is seldom enough rain in these arid climates to carry the Calcium any deeper, it often forms a hard, whitish-colored layer that is impervious to water, known as caliche. This caliche layer is akin to concrete and must be broken up to allow deep-rooted plants such as fruit trees to extend their roots into the subsoil and to allow drainage of irrigation water.

Agricultural Sulfur (90-100% S) must be converted to the sulfate (SO_4) form before it can be used by plants. This is done by Sulfur converting bacteria which naturally occur in soils. The soil temperature must be above 13°C (55°F) for the bacteria to do their work. During the conversion, 4 atoms of Oxygen will be taken from H_2O to form SO_4, leaving 8 atoms of free Hydrogen H+, which is acidic and will lower the soil pH. When Sulfur is applied as Ca, Mg, or K sulfate it generally will not affect the soil pH much. Ferrous sulfate (Iron sulfate) will lower the soil pH.

On the next page is the soil report that we are working with, showing the requirements for Calcium, Magnesium, Potassium and Sodium that we have determined in this chapter.

Soil Report and Comments

Element		Results	Comments
Total Cation Exchange Capacity TCEC		11.4	
pH of Soil Sample		5.58	
Organic Matter %		5.6%	
Anions			
Sulfur S (parts per million ppm)		20	
Phosphorus P ppm		100	
Cations			
Calcium Ca++ ppm	Desired Found Deficit	1550 1250 -300	Add 300 ppm Calcium Ca
Ca Base Saturation 60-70 %		55.0%	
Magnesium Mg++ ppm	Desired Found Deficit	164 116 -48	Add 48 ppm Magnesium Mg
Mg Base Saturation 10-20 %		8.5%	
Potassium K+ ppm	Desired Found Deficit	178 89 -89	Add 89 ppm Potassium K
K Base Saturation 2-5 %		2.0%	
Sodium Na+ ppm	Desired Found Deficit	39 26 -13	Add 13 ppm Sodium Na
Na Base Saturation 1-5 %		1.0%	
Other Bases		6.2%	
H+ Exch Hydrogen 10-15%		27.0%	
Other Elements ppm			
Boron		0.21	
Iron Fe		50	
Manganese Mn		11	
Copper Cu		1.07	
Zinc Zn		16.4	
Aluminum		1841	Normal

A printable periodic table of the elements in .PDF form may be found here:
http://www.webelements.com/nexus/sites/default/files/webelements_table_5sf_2012-06-07.pdf

Chapter 4

The Primary Anions
Phosphorus P, Sulfur S, and Chlorine Cl

In chapter 3 we looked at some soil test results and calculated how much of the various **primary cation** nutrients needed to be added to the soil to bring the base saturation level to our ideal soil balance of 68% Ca, 12% Mg, 4% K, and 1.5% Na.

In this chapter we will learn to calculate the required amounts of the **primary anions** Sulfur, Phosphorus, and Chlorine

Here is what we calculated for the Ideal Soil ratio of cations in chapter 3:

Total of Major Cations after Balancing

Calcium	Magnesium	Potassium	Sodium
1550 ppm	164 ppm	178 ppm	39 ppm

All of the rest of our calculations for the soil mineral prescription will be based on these numbers.

Phosphorus P and Potassium

Here is the Phosphorus section of the Ideal Soil chart:

Other major nutrients (anions)

Phosphorus P- min 100 ppm	P = **Ideal** K by weight (ppm) **BUT: phosphate** (P_2O_5) should be **~2X potash** (K_2O)	Needs a highly bio-active soil to keep it available.

The chart says that Potassium and Phosphorus should be equal by weight in the soil. If we are going to end up with 178 ppm of Potassium; we want to end up with 178 ppm of Phosphorus too. This gets a little tricky for a couple of reasons:

- On a fertilizer label in the USA and some other countries, the letters N, P, and K do not stand for Nitrogen, Phosphorus, and Potassium. They stand for Nitrogen, phosphate, and potash. N is elemental Nitrogen, but phosphate is P_2O_5, 2 atoms of Phosphorus and 5 atoms of Oxygen. Potash is K_2O, 2 atoms of Potassium and 1 atom of Oxygen. More on this below, but in effect the P on the fertilizer label is only 44% Phosphorus by weight, while the K on the fertilizer label is 83% Potassium by weight. There is almost twice as much Potassium by weight in potash as there is Phosphorus in phosphate. This is why the Ideal Soil chart says **phosphate should be 2X potash**.

- Different soil testing laboratories use different conventions when reporting results. Some labs report P as actual Phosphorus, some report it as phosphate. It is the same with Potassium and potash. Some lab soil reports list Potassium in ppm K, but show Phosphorus as P_2O_5 phosphate. Whatever lab one is using, it is necessary to know what form of P and K they are reporting.

If you are buying a commercial NPK labeled fertilizer in the USA, and you want to maintain the Ideal Soil ratio of P=K by weight, you would not want one labeled 10-10-10, but rather one labeled 10-20-10, or 5-10-5. Let's look at that a little closer:

Say you bought an NPK fertilizer that was labeled 10-10-10, and knew that meant 10% Nitrogen, 10% phosphate, and 10% potash. 10% is elemental Nitrogen N, but the phosphate is only 44% Phosphorus and the potash is only 83% Potassium. The actual ratio and amount of elemental nutrients in that 10-10-10 is 10% N, 4.4%P, and 8.3%K. A label that said 10-20-10 would be 10%N, 8.8%P, and 8.3%K, much closer to the P=K ratio by weight that we have called for in the Ideal Soil. This writer has read different stories as to why the P and K on fertilizer labels are listed as they are, one theory being that early analytical chemists only purified the samples to the oxide form and then weighed that. The best guess as to why they are *still* listed that way is that it makes it appear there is more fertilizer in the bag than there actually is.

Here's why phosphate is 44% actual Phosphorus:

The chemical formula of **phosphate is P_2O_5,** 2 parts Phosphorus and 5 parts Oxygen. The periodic table of the elements tells us that the atomic weight of Phosphorus is 31, and that of Oxygen is 16. So we have

2 x 31 = 62 (the weight of P)
5 x 16 = 80 (the weight of O)

added together, the weight is 142. Divide the weight of P by the total weight

62 / 142 = 0.437, or 44% P

For potash, the formula is K_2O, 2 parts Potassium and one part Oxygen. Potassium's atomic weight is 39, Oxygen is still 16.

2 x 39 = 78 (the weight of K)
1 x 16 = 16 (the weight of O)

The total weight is 94. Divide the weight of K by the total weight

78 / 94 = 0.829 or 83% actual elemental K.

If the soil test report you are working from lists Phosphorus as P_2O_5, you should first convert P_2O_5 to elemental Phosphorus by multiplying P_2O_5 x 0.44.

Calculating Phosphorus

Figuring out the amount of Phosphorus we need is simple and straightforward. Here is the Anion section from the soil report:

Anions	
Sulfur S (parts per million ppm)	20
Phosphorus P ppm [227 ppm P_2O_5]	100

Our Potassium (actual K) is 178 ppm, so we simply subtract the amount of actual P shown on the soil report from the amount of total K in our Ideal Soil
178 ppm – 100 ppm = 78 ppm

We will need to add 78 ppm of elemental Phosphorus to the soil.

Sulfur S

Here is the Sulfur part of the Ideal Soil chart:

Sulfur S - - min 50 ppm	1/2 x Ideal K up to 300 ppm	Need for Sulfur amino acids Conserves soil N and Carbon.

Here is the Sulfur reading from our soil report:

Sulfur S (parts per million)	20

We calculated above that we wish to end up with 178 ppm of Potassium K, and the Ideal Soil chart tells us that we want ½ as much Sulfur as "Ideal" K (K=4% of CEC), so we divide the K amount by 2:

178 / 2 = 89 ppm S would be our desired amount.

Subtract our existing Sulfur level of 20 ppm:

89 - 20 = 69 ppm

We need to add **69 ppm** of Sulfur S.

Chlorine Cl and Potassium Chloride KCl Fertilizer

Chlorine is an essential nutrient for plants, animals, and people, but it is not measured on a standard soil test. Under natural conditions Chlorine in supplied from the breakdown of chloride minerals in the soil such as Sodium or Potassium chloride. Cl is also naturally found in rainwater, especially near seacoasts. Areas far from the ocean such as mid-continent areas may have little natural Cl in the atmosphere or rain but may still have significant amounts derived from burning fossil fuels.

Chlorine deficiency is rare in most agricultural soils today; chlorine excess is much more common, due to using chlorinated water for irrigation or even more commonly, Potassium chloride fertilizers. Most commercial fertilizers that contain Potassium are formulated with Potassium chloride, KCl, because it is cheap and readily available. It is also effective in the short term, giving a strong growth response. However, in large amounts it is toxic to soil organisms, from bacteria to earthworms. Use of KCl will also prematurely age the clay in the soil, reducing the exchange capacity.

Another major drawback is that KCl fertilizer can rapidly deplete the upper soil layers of Calcium. In the soil the K+ can exchange for Ca++ on a negative – site (or KCl can react with free Ca or with lime present in the soil), releasing a Calcium ion, but that Ca++ ion may immediately bond with the Chlorine ion from the KCl, forming Calcium chloride $CaCl_2$. Calcium chloride is highly water soluble and will easily leach to a lower soil horizon.

It is impossible to have a healthy, living soil if Potassium chloride is being used in large quantities, e.g. more than 20% of K supply. Potassium sulfate, while more expensive, does not harm the soil life. Potassium sulfate is allowed for Certified Organic use by the USDA National Organic Program. Potassium chloride is not allowed.

From the Ideal Soil Chart

Chlorine (Cl)- min 25ppm	1x to 2x Sodium	Essential, but ages clays rapidly when used in large amounts

The simplest way to ensure adequate Chlorine in the soil is to add regular salt, Sodium chloride NaCl. Sea salt, mineral salts like Redmond's, or even table salt. NaCl is 40% Sodium and 60% Chlorine.

If there is reason to believe the soil needs more Chlorine, but the soil already contains adequate or high levels of Na, or if amendments like Sodium nitrate are to be used, Potassium, Magnesium, or Calcium chloride could be used as a Cl source instead of NaCl.

Amount of Primary Cations and Anions needed

These are the primary plant nutrients that are measured on a standard soil test. In the next chapter we will discuss the minerals Boron, Iron, Manganese, Copper

Anions			
Sulfur S (parts per million ppm)		20	Add 69 ppm Sulfur S
Phosphorus P ppm		100	Add 78 ppm Phosphorus P
Cations			
Calcium Ca++ ppm	Desired Found Deficit	1550 1250 -300	Add 300 ppm Calcium Ca
Ca Base Saturation 60-70 %		55.0%	
Magnesium Mg++ ppm	Desired Found Deficit	164 116 -48	Add 48 ppm Magnesium Mg
Mg Base Saturation 10-20 %		8.5%	
Potassium K+ ppm	Desired Found Deficit	178 89 -89	Add 89 ppm Potassium K
K Base Saturation 2-5 %		2.0%	
Sodium Na+ ppm	Desired Found Deficit	39 26 -13	Add 13 ppm Sodium Na
Na Base Saturation 1-5 %		1.0%	

and Zinc.

A printable periodic table of the elements in .PDF form may be found here:
http://www.webelements.com/nexus/sites/default/files/webelements_table_5sf_2012-06-07.pdf

Chapter 5

The Secondary Elements
Boron, Iron, Manganese, Copper and Zinc

These five will complete our analysis of the nutrient minerals that are measured on a standard soil test.

B Boron
Fe Iron
Mn Manganese
Cu Copper
Zn Zinc

Here is the part of the soil report we will be working with in this chapter:

Other Elements ppm	
Boron B	0.21
Iron Fe	50
Manganese Mn	11
Copper Cu	1.07
Zinc Zn	16.4

Boron B

Boron is one of the more rare elements on planet Earth. On average, the earth's crust contains 50,000 ppm of Calcium but only 9 ppm of Boron.

Boron is only mined in a few dry places. Turkey and the Mojave Desert of California are the world's primary sources of Boron. Some boron is also refined from the Sodium nitrate deposits found in the Atacama desert of Chile.

Boron is a close partner with Calcium; Calcium transports many nutrients into the plant and within the plant, but it needs boron to keep it mobile. The saying is that Calcium is the truck, Boron is the driver.

Free Boron in the soil is highly water soluble and leaches out easily. Plants also take it up readily. In most high-production agricultural soils Boron (and Sulfur) will need to be applied every year. Boron seems to be held on the organic (Carbon) portion of the soil and is only available to the plant when soil moisture is adequate in the upper organic horizon of the soil. Alfalfa crops will often show a Boron deficiency in mid to late summer if the topsoil gets very dry; the deep rooted alfalfa can still bring up water from the subsoil, but the Boron is tied up in the organic matter in the dry topsoil. If the field cannot be irrigated, it may be effective to apply a foliar Boron spray. This spray should be highly diluted to no more than 1 or 2 pounds of Boron per acre (1 or 2 kg of B per hectare).

Although absolutely essential, Boron can also be toxic to soil life in high doses and is known to inhibit the sprouting of seeds. As long as one keeps it close to the ratio of 1 part Boron to 1000 parts Calcium there will be no problems.

Here is the Boron section from the Ideal Soil chart:

Boron B – min 1 ppm	1/1000 of Calcium (max 4 ppm)	Essential for Calcium utilization.

Simple enough. 1/1000th of Calcium. Our desired Calcium level is

Calcium Ca++ ppm	Desired	1550

1550 ppm / 1000 = **1.55 ppm** Boron desired

Our Boron level from the soil report is

Boron B-	0.21

Subtract the soil test reading from the desired level

1.55 - 0.21 = 1.34 ppm of B

We need to add 1.34 ppm Boron.

Iron Fe

From the Ideal Soil Chart:

Iron(Fe) + min 50ppm Manganese(Mn) + min 25ppm Zinc (Zn) + min 10ppm Copper (Cu) + min 5ppm	Fe: 1/3 to 1/2 x **Ideal** K Mn: 1/3 to 1/2 x Fe Zn: 1/10 x P (up to 50ppm) Cu: 1/2 x Zn (up to 25ppm)	Iron and Manganese are twins/opposites and synergists, as are Copper and Zinc.

Iron should be 1/3 to 1/2 of Ideal K, Potassium. We decide to set the desired Fe level at 1/2 of K, the same level we used for Sulfur S. Desired K (4% of CEC) is 178 ppm; we will want to end up with 1/2 of that:

178 ppm / 2 = **89 ppm total Fe desired**

The soil report we are using tells us we have

Iron Fe	50

89 ppm – 50 ppm = 39 ppm

39 ppm of Iron should be added to this soil.

Manganese Mn

Manganese is essential for the production of fertile seeds. There is an atom of Manganese at the center of the germ of every seed. Fruits like peaches and plums are often found with a shriveled seed if the plant is deficient in Manganese. This does not necessarily mean the soil is lacking in Manganese; it may simply be deficient in one of the other essential minerals such as Iron, Zinc, or Copper. They all need to be there in their proper proportion.

The Ideal Soil chart calls for **Mn to be 1/3 to 1/2 of Iron**. Unless the soil CEC is above 15 meq and the test shows it contains above 150 ppm Fe, we do not need or want to go above 50 ppm Manganese, and ideally, we do not want Mn to be more than ½ of Iron. One reason for this is that high levels of Manganese have been linked to BSE (Mad Cow Disease) and other degenerative neurological ailments, especially in soils that are deficient in Copper and Zinc. This does not mean that all high Mn soils are dangerous or will pose problems, but it is wise to add sufficient Iron to a high Mn soil. See the section on Manganese in the appendix for more details.

The soil report tells us that the Mn level is

Manganese Mn	11

Our desired **Iron** level is 89 ppm, so "ideal" Manganese should be ½ of that or 45 ppm.

45 ppm – 11 ppm = 34 ppm

34 ppm of Manganese should be added to this soil.

Copper Cu

Copper should equal ½ of Zinc. Zinc is 16.4 ppm, which we will decide below is adequate. Copper needs to be

16.4 / 2 = **8.2 ppm total desired Copper**

The soil report reads

Copper Cu	1.07

8.2 ppm - 1.07 ppm = 7.13 ppm Copper needed

We need to add 7.13 ppm of Copper to this soil.

Zinc Zn

The Ideal Soil chart says Zinc should equal 1/10 of Phosphorus. In the last chapter we determined our ideal Phosphorus level for this soil to be 178 ppm (100 ppm in the soil, plus 78 ppm to be added).

178 ppm x 0.10 = **17.8 ppm** total is our desired level of Zn.

The soil report reading for Zinc is 16.4ppm

17.8 ppm – 16.4 ppm = 1.4 ppm. This is less than 10% of the ideal total, and well within the margin of error for soil sampling and lab testing. It is close enough.

We do not need to add any Zinc.

That's it. We are done calculating the amounts of the primary and secondary minerals that need to be amended. Next we will take a brief look at the trace minerals and micro-elements, and then we will be ready to write the soil prescription.

The completed soil report with all of our work so far is on the page following the comments about Copper and Zinc below

Notes on Fertilizing with Copper and Zinc:
The Ideal Soil guidelines are for elemental Zinc to be at 10% of elemental Phosphorus by weight, and for Copper to be 1/2 of Zinc.

Caution is advised when amending the Copper levels in the soil because "free" Copper can be toxic to soil organisms as well as fish and other aquatic life. Keep Copper out of streams and ponds; don't apply close to the water or where the Cu can wash into to the water before it soaks into the soil.

As a general rule, it is safe enough to add 3.5 to 5 ppm of elemental Copper to most soils at any one time. Soils that are high in organic matter, high CEC clays with a good amount of Calcium, and high Ca soils in general have a large buffering capacity and can easily adjust to higher amounts of Copper in a single application than low CEC soils.

The Copper example above calls for 14 kg/ha or 7ppm of Copper. This is a safe amount to add to this soil in a single application as long as the other minerals on the soil Rx are added as well.

Sheep need copper but more than a little can be deadly to them, causing liver damage. Once the added Copper has been assimilated into a mineral balanced, biologically active soil the plants should not take up excessive Copper that might pose a danger to sheep. In addition whatever Copper the forage does contain will be balanced by the proper amount of many other minerals. Nonetheless, one will want to err on the low side when prescribing Copper for sheep pastures, and not add a large amount at once.

2.5 ppm Cu would be a safe amount for a single application when amending pastures for sheep or other grazing animals that are known to be Copper-sensitive. An interesting way of learning if the pasture or feed is Copper deficient is to wire a piece of copper tubing to a fence or other area where the animals spend time. If the animals lick the copper tube enough to make it shiny, they need copper.

Free Copper can also be hard on fungi and on the photosynthesizing algae that grow on the soil surface. The safest way to apply Copper only (as a solo amendment) is to mix it with some organic matter such as compost.

Zinc: The amount of Zinc that can be added to agricultural soils per year is regulated by law in some parts of Canada and the USA; in the US state of Washington the limit is 7 lbs of Zinc per acre per year (14 kg/ha or 3.5ppm added to the top 15 cm). Oddly enough, Copper as a fertilizer is not regulated in Washington. It is our understanding that the Washington State rules were copied and pasted from pre-existing Canadian rules.

For a more in-depth discussion of Copper and Zinc, see the appendix.

Soil Report and Comments

The completed worksheet, in ppm, for chapters 3 through 6 of The Ideal Soil v2.0

Element		Results	Comments
Cation Exchange Capacity CEC meq		11.4	
pH of Soil Sample		5.58	
Organic Matter %		5.6%	
Anions			
Sulfur S (parts per million ppm)		20	Add 69 ppm Sulfur S
Phosphorus P ppm		100	Add 78 ppm Phosphorus P
Cations			
Calcium Ca++ ppm	Desired Found Deficit	1550 1250 -300	Add 300 ppm Calcium Ca
Ca Base Saturation 60-70 %		55.0%	
Magnesium Mg++ ppm	Desired Found Deficit	164 116 -48	Add 48 ppm Magnesium Mg
Mg Base Saturation 10-20 %		8.5%	
Potassium K+ ppm	Desired Found Deficit	178 89 -89	Add 89 ppm Potassium K
K Base Saturation 2-5 %		2.0%	
Sodium Na+ ppm	Desired Found Deficit	39 26 -13	Add 13 ppm Sodium Na
Na Base Saturation 1-5 %		1.0%	
Other Bases		6.2%	
H+ Exch Hydrogen 10-15%		27.0%	
Other Elements ppm			
Boron B		0.21	Add 1.34 ppm Boron B
Iron Fe		50	Add 39 ppm Iron Fe
Manganese Mn		11	Add 34 ppm Manganese Mn
Copper Cu		1.07	Add 7.13 ppm Copper Cu
Zinc Zn		16.4	OK
Aluminum		1841	Normal

A printable periodic table of the elements in .PDF form may be found here:
http://www.webelements.com/nexus/sites/default/files/webelements_table_5sf_2012-06-07.pdf

Chapter 6

Micro Elements and Nitrogen
(With a few notes on Beneficial Soil Organisms and Humates)

The bottom section of The Ideal Soil chart:
Micro (trace) Elements

Chromium Cr- Cobalt Co+ Iodine I- Molybdenum Mo- Selenium Se- Tin Sn+ Vanadium V+ Nickel Ni+ Fluorine F–	All of these are essential in small amounts. 0.5 - 2ppm is enough. Some of the micro elements (e.g. Mo, Se) can be toxic to plants and soil organisms in quantities above 1-2ppm. Use Caution when applying micro/trace elements in purified forms	There are probably 30 or so other elements needed in a perfect soil. Sources are amendments such as *seaweed, rock dust, ancient seabed or volcanic deposits, rock phosphate greensand* etc.

The science of micro (trace) minerals and their relationship to soil, plant, and animal health is still in its infancy. Until a few years ago no one had any idea that Chromium and Vanadium were essential nutrients but they assuredly are.

At soilminerals.com we rely mostly on natural micro mineral sources such as those listed in the right-hand column of the table above; these sources contain dozens or scores of different elements. Most micro minerals are only needed in very tiny quantities; often a few parts per billion are sufficient.

A standard soil test does not test for these micro nutrients. It tests only for the elements we have examined in the previous chapters, with the addition of perhaps Nitrogen or Aluminum. Any chemical assay only measures the elements that are specifically being tested for.

Most soil testing laboratories will do special tests, such as for Cobalt, Molybdenum, or Selenium, but these are individual tests that must be requested and paid for in addition to the cost of the standard test. An example of where it may be worthwhile to request a micro element test would be if one had a large area of pasture and suspected that an essential trace mineral like Selenium or Cobalt was seriously deficient. If it is not practical or affordable to apply hundreds of pounds per acre of a broad-spectrum micro mineral source to a large area, it may be practical to apply a few ounces per acre of a refined source.

The element **Selenium** is a good example of this. Selenium is essential for proper immune function, it is a co-factor with vitamin E, and it is associated with resistance to viruses. Severe Selenium deficiency in pasture and feed leads to white muscle disease in ruminants, which is fatal. In the 1970s investigations into high rates of heart disease in the Keshan province of China led to the discovery of the essential role of Selenium in human nutrition. The sandstone bedrock that had formed the basis of the soil in Keshan province was found to be completely lacking in Selenium, and that lack turned out to be the source of the ongoing epidemic heart disease. Prior to this discovery, Selenium was considered a toxin, which it can be in concentrations greater than a few parts per million.

Adding Selenium to the soils of Keshan province produced a dramatic decline in heart disease. The country of Finland, which also had very high rates of heart disease, turned out to also be Selenium deficient, and reportedly all of the agricultural land in Finland was subsequently amended with Selenium.

The so-called locoweed of the American West is a member of the pea family that, when growing on high-Selenium soil, concentrates enough Selenium to be toxic to animals grazing on it.

Cobalt is an essential mineral that is only needed in minute quantities. Ruminant animals such as cattle and sheep produce vitamin B-12 in their digestive system when Cobalt is available in their forage. Without Cobalt they are subject to hookworm, bacterial infections, brucellosis, and neurological diseases. This connection was first made in cattle on Cobalt deficient pastures in Florida, USA, during the 1930s and has since been found to be a problem in many areas of the world ranging from Russia to New Zealand. As little as two ounces of Cobalt per acre can alleviate the deficiency.

Molybdenum: All of the known Nitrogen-fixing bacteria require Molybdenum as a catalyst, both the free-living types and those such as rhizobia that form nodules on the roots of legumes. No Molybdenum, no Nitrogen fixation in the soil. Only vanishingly small quantities of Molybdenum are needed.

We recommend wherever possible that natural sources of multiple trace minerals be used, in addition to whatever primary and secondary minerals have been found lacking on the soil test. Kelp and sea salt, as well as ancient sea bed deposits, have the widest range of trace minerals to be found anywhere. Regular ocean salt contains at least 80 different minerals.

Sea salt and Mineral Salts
Many people are fearful of using **sea salt** on their gardens or fields. This fear may go back to the ancient story of the Romans salting the fields of Carthage after the Carthaginians were defeated. Whether there is any truth to that tale, who knows, and who knows what type of “salt” was used? While it is true that high concentrations of Sodium are harmful to plant and soil life, many inland soils are Sodium deficient, especially in high rainfall areas. After the deadly tsunamis that washed over the shores of the Indian Ocean in 2004, many expected the inundated farmlands to be harmed, but all the reports this writer has seen indicate that the subsequent crops were excellent. We generally try to keep Sodium

concentrations below 3% base saturation, but anything up to 5% is no problem for most food crops, and in general levels up to 10% of CEC will not pose a problem as long as the other primary cation minerals are in balance . Sea salt is at most a little over 40% Sodium. In the 11.4 meq soil report that we are using as an example in this book, 5% base saturation would be over 650 kg/ha of sea salt, or 6.5kg per 100 square meters. Sugar beet, celery, Swiss chard, turnips, and spinach are crops that benefit from relatively high Sodium levels.

Redmond mineral salt is a popular culinary, livestock feed, and soil amendment salt from the USA. Redmond salt is mined from an ancient seabed in the state of Utah. It contains over fifty minerals.

Kelp and other seaweeds are some of the very best micro mineral sources, and some of the few good sources of **Iodine**. Sea water contains a small amount of Iodine, but sea salt has been evaporated either in the hot sun or an oven. Iodine being very volatile, most of the Iodine is lost in the drying process. Seaweeds concentrate Iodine and lock it in as part of their organic structure. Seaweeds also contain interesting plant hormones that act as growth stimulants and many amino acids to feed the soil life and the plants.

The only significant mineral source of **Iodine** is unrefined **Chilean Sodium nitrate**, which contains 0.04-0.08% iodate as well as being a good source of **Boron** and highly available nitrate Nitrogen.. USDA NOP organic rules allow **Chile nitrate of soda** to be used for up to 20% of yearly N application. Whether or not the presumably refined Chile nitrate being marketed in the US in 2013 contains significant amounts of Boron and Iodine is presently unknown.

Rock Dust

The simplest solution for micro minerals in the soil is to find a local rock or gravel quarry that has a crusher operation and make a deal with them for their crusher dust. It is a waste product to them so expect to get a good price. One should do their best to determine the mineral makeup of the rock dust; if the quarry has contracts with the government or large construction firms they may have been required to get a chemical assay of the rock and will know the mineral composition of their product.

Another facet of newly crushed rock is that it is sharp; the crystals are freshly broken and have sharp points and edges. Energy, heat, and electric charge all go to a point, which could conceivably increase electrical charge in the soil; a good thing in many soils, especially older soils, clay soils, and those very high in organic matter such as peat and muck soils. In addition, sharp edges and points make it easier for soil acids and microbes to attack the rock and etch new, fresh minerals from them.

Freshly ground rock may also increase the paramagnetic qualities of the soil. **Paramagnetism** is a rather new aspect of agriculture having to do with a soil's ability to interact with the Earth's magnetic and electrical field. The author has done trials with highly paramagnetic basalt from Canada with very good results. Many soils, especially older agricultural soils, are deficient in paramagnetism. A detailed discussion of the subject is beyond the scope of this chapter, but those intrigued should check out the work of Phil Callahan PhD, starting with his book *Paramagnetism*. Oxygen is the most paramagnetic element of all, and simply getting Oxygen into your soil will go a long way toward increasing its energy level.

Glacial rock dust does not come from glaciers, at least not directly. The big buzz on glacial rock dust began with the publication of John Hamaker and Donald Weaver's book *The Survival of Civilization* in 1982. Hamaker and Weaver argued that the last time the planet's soils had a good dose of fresh minerals was when the glaciers melted at the end of the last ice age, around 10,000 years ago, and deposited the loads of boulders, gravel, and fine rock dust that they had picked up while moving toward the equator from polar regions. The popularity of glacial rock dust is no doubt also due to Dr. Robert McCarrison's 1921 book *Studies in Deficiency Diseases*, where he described the long-lived and healthy people of the Hunza valley in the Himalayas. Their fields and gardens were irrigated from mountain streams running off of glaciers, and the water was milky-colored from the amount of rock dust suspended in it.

What is marketed today as glacial rock dust is crusher dust from a quarry, but the quarry is located on or in a glacial till or moraine, a large deposit of rocks left behind when the glaciers retreated. Once again it is generally a waste product from the crusher operation and should not be terribly expensive. The advantage of so-called glacial rock dust is that the moraine consists of a mixture of rocks, some perhaps carried from hundreds of miles away, and will have a wider range of minerals than the dust from a quarry crushing local bedrock.

More recommended micro mineral sources:

Azomite Volcanic Rock Powder: An ancient deposit of volcanic ash that later became a sea bed, Azomite is a superb source of 67 naturally chelated minerals. Azomite stands for "A to Z Of Minerals Including Trace Elements." Azomite is mined from the "pink" hills of Utah.

Tennessee Brown Phosphate: Tennessee brown phosphate is highly reactive and highly available, and is a good choice for soils above pH7. Total phosphate content runs from 21 to 25%. Tennessee Brown phosphate contains more than 50 micro elements.

Colloidal Clay Phosphate, soft rock phosphate, SRP, CalPhos: A soft, powdery phosphate, Calcium, and micro element source from Florida. High exchange capacity and readily available in acid soils. An average analysis for this amendment (reproduced in the appendix of this book) lists 27 minerals.

Jersey Greensand: This famous slow release Potassium and Iron source is from an ancient seabed in New Jersey. J. I. Rodale recommended it highly. It is also a source of Calcium, Magnesium, and phosphate along with 30 or more micro elements. Greensand, also called glauconite, is found in many parts of the world, not just New Jersey. There may be a glauconite deposit close to you. The mineral ratio in the glauconites can vary quite a bit.

Humate Ores and Humic Acid Sources

Whenever available, we recommend adding a source of humic and fulvic acids to the soil. These amendments stimulate soil life, increase exchange capacity, and bring life and oxygen to the lower levels of the root zone, which increases the friability of tight soils and creates deeper topsoil. The most common sources are humic shale, lignite coal (a soft peat coal also known as Leonardite), and humate deposits. Humate deposits come from the remains of plant and animal life that accumulated in an ancient freshwater sea in the western US and other similar locations around the world. Humate ores are frequently good sources of micro minerals and often contain high levels of Potassium.

Does your soil need micro minerals?

Whether or not you need to add a micro element source to your soil depends on what type of soil you have and its past history. A dense tropical or subtropical clay that has been leached and weathered for millions of years is unlikely to contain many micro elements. Such a soil is very likely in need of all the help it can get. A chernozem type prairie soil, such as found in the upper-central USA, may have plenty of micro mineral reserves, especially if the area was glaciated and contains jumbled rocks from many sources. A coarse and undeveloped soil, the type found in many arid regions, may or may not need additional micro minerals, depending on the source rocks. In all cases, creating a biologically active soil and bringing the humus level up to optimum will not only help to make whatever minerals are present more available, but also help to retain them in the root zone.

Another consideration is what the soil has been used for in the past. Long-term agricultural cropping of soils where the minerals have been taken up by the plants and the crops harvested and taken away depletes all of the mineral reserves.

It is possible to assay all of the minerals in a sample of soil, including those that are still in the form of sand, gravel, and rocks. This is done by grinding the soil to a very fine powder and dissolving it completely in a heated solution of Aqua Regia, a mixture of concentrated ntric and hydrochloric acids. From this assay one could determine the total amount of, for example, Calcium that the root zone contained, measure the amount of Calcium that the crops being grown took up each year, and calculate how many years a crop could be grown on that soil before the amount of Calcium or other mineral nutrient had been completely taken up and shipped away.

We have found one chart showing a complete assay of soil minerals and an estimate of how many years' supply the soil contains. The table was published in *Soil Chemistry* by Bohn, McNeal, and O'Connor in 1985. It is based on the work of the scientist Vinogradov whom we assume was Russian. Here are a few examples from that chart:

Total Soil Mineral Reserves to 1 Meter Depth

Mineral element	Plant uptake per year kg/hectare	Years of supply at average plant uptake
Calcium	50	260
Potassium	30	430
Magnesium	4	4,600
Iron	0.5	100,000
Sulfur	2	320
Selenium	0.0003	40

According to the notes on the original, these are the total reserves in the soil to a depth of 1 meter. No indication is given of where this soil was from, but clearly it was a fertile soil if it contained potential reserves of Potassium to feed crops for 430 years. The question of which crops is not addressed.

Looking at the Sulfur reserves, 320 years at 2 kg per year would give us 640 kg/ha of Sulfur in this soil.

Perhaps some crops may only take away 2 kg/ha (1.8 lbs) of S per year from the soil, but more commonly crops will uptake 20 or more kg/ha. A good crop of sugar beets will use as much as 50 kg/hectare per year of Sulfur. That 640 kg of Sulfur would only be enough to grow sugar beets for a dozen years, even assuming the plants could somehow access every bit of Sulfur in the top 100cm of soil (which of course they couldn't).

We can see how a soil could be rapidly depleted.

An even more dramatic example is Selenium in the bottom row of the chart-- only enough for 40 years of growing, and as our arithmetic above has shown, perhaps for much less time than that.

We begin to see why soils are spoken of as "worn out". This can be the consequence of short-sighted exploitative agriculture where soluble NPK fertilizers have been applied in large amounts in order to force maximum growth and yield. Plants cannot live and reproduce solely on a diet of NPK; they must have the other essential minerals too, so they draw them from the soil reserves, depleting those reserves year after year. As the crops are harvested and sold away; the minerals are not replaced. In effect, this is mining the soil. A mine is only a paying proposition as long as the mineral being mined lasts.

Many agricultural soils worldwide have been abused over the last few thousand years. The plains of North Africa that once were the breadbasket of the Roman Empire are now deserts where even goats can barely survive. This situation has not improved since soluble NPK fertilizers were introduced in the 1800s.

The following two paragraphs from the end of chapter 1 are copied here because they bear repeating:

If we look at agricultural soils from a nutritional standpoint, they are much more than an anchor for the roots, a base to keep the crops from falling over. Each crop harvested and taken away depletes the soil's store of essential nutrient minerals. If the minerals are not replaced, we eventually reach a point where there are not enough left to grow a healthy crop with the ability to mature seeds for the next generation. Long before this point is reached, the nutrient density of the crop for human and animal food has suffered. Much of our arable land worldwide is producing empty calories, mostly carbohydrates made from the atmospheric elements Carbon, Hydrogen, and Oxygen. The solution, the only solution (barring the ability of plants or soil organisms to transmute elements alchemically), is to supply these needed minerals from a source where they are abundant. That source should ideally be located as close as possible to where the minerals are needed in order to minimize transportation costs. It makes no sense to ship ground limestone across the country when every state in the USA has limestone deposits, but when it comes to rare elements like Selenium or Boron which are only found in concentrated form a few places in the world, the transport costs are justified.

Mining of the needed minerals need not entail long-term environmental damage either. Mines and quarries can be carefully worked by those who care about their home planet, and when the mines are depleted they can be landscaped and planted to be as or more beautiful than before mining. It's also worth noting that many of the economically viable sources for agricultural minerals contain such high concentrations of these minerals that they are toxic to soil life and little or nothing grows there. Removing these toxic concentrations and using them to make other parts of the planet healthier and more productive can, at the same time, open up these formerly toxic soils to the growth of forest or grasslands. None of this should be done on the basis of greed or short-term gain, but rather wisely, intelligently, and in harmony with Nature.

Beneficial Bacteria and Fungi

BSOs: Beneficial Soil Organisms

We recommend and use beneficial soil fungi and bacteria for the same reason that bakers use bread yeast, champagne makers use champagne yeast, brewers use beer yeast, yogurt makers use yogurt starter, and cheese makers use the culture that will make the kind of cheese they wish to make. One can wait around and hope that the right yeast spore will fall into the vat and create champagne, or one can increase the odds in their favor by adding the cultures that they want right at the start. We want to be sure that the best possible soil life "culture" is there and ready to make the minerals and organic nutrients in the soil available.

Beneficial Soil Organisms (BSOs) have proven their value in bringing vibrant life to the soil, in protecting the soil and plants from disease and drought, and in making soil nutrients readily available.

BSOs are also valuable for freeing nutrients that are in the soil but "tied up". For example, soils may have good phosphate reserves, yet the plants growing on those soils are starving for phosphate. Many farmers and gardeners in this situation resort to adding soluble phosphate fertilizers to a soil that already has plenty of phosphate, only to have the added phosphate chemically combine with cations in the soil and become unavailable to plants within a few weeks. The reason for this is that elemental Phosphorus is an extremely active "acid" mineral. It has a powerful negative - charge, and just can't wait to latch onto the nearest + charge. The nearest + charge is usually Calcium, and together Ca and P form Calcium phosphate, a very stable compound and exactly what bones are made of. Seen any water-soluble bones lately? It takes either strong acids or the right microorganisms to break the Ca-P bond.

Another good reason to use pro-biotic soil cultures is that many soils have a hard time breaking down organic matter to form humus. Seeding the soil or the compost pile with the right biology will greatly assist in breaking down tough plant roots, wood chips, corn stalks, and other "chunks" of organic matter in the soil, releasing the tied-up nutrients all along the way.

Here is a third important reason to use a good BSO culture to seed the soil: Healthy soils always contain healthy symbiotic fungi. 95% of plant species have a natural and ancient alliance with the soil fungi. Some fungi send out their long "roots" called hyphae (hi-fee) for many meters, both deep into the soil and just below the surface in the topsoil. They search out nutrients and moisture and bring them back to share with the plants. Other fungi are smaller, and only live close to the plants they are partners with. They break down small rock particles while searching for nutrients, digest dead organic matter, and even protect the plants from disease by producing antibiotics; the originals for all of our antibiotics were from the fungi. (Penicillin is produced by the common blue bread mold.) In exchange for being fed, helped, and defended, the plants feed the fungi, sharing the sugars and other carbohydrates that they make in the open air and sunshine with the fungi that never see the sun.

The name for these types of symbiotic fungi is mycorrhizae (my-ko-riz-ee), or mycorrhizal (my-ko-riz-al) fungi. Myco means fungus and rhizae comes from rhizome and means roots. Fungus root or root fungi is what they're named and what they are.

Some mycorrhizae (my-ko-riz-ee) actually intergrow with the tiny root hairs of the plants. The plant's roots are then a combination of fungus and root tissue; the fungi share their own sap with the plants that are their partners; this type is called endo-mycorrhizae because it lives partly within (endo=inside) the plants that are its symbiotes. 90% of cultivated plants partner with endo-mycorrhizae. Other mycorrhizae grow right next to the plant roots, sharing nutrients with the plants through the soil water/nutrient solution. Ecto- means outside; ecto-Mycorrhizae live entirely outside the plant roots. Most trees and shrubs partner with ecto-mycorrhizae. A good tip if you are planting new or having trouble with old shrubs and trees, either ornamentals or fruiting types, is to find a place where the same plant is healthy and growing well and scrape off a little of the topsoil or surface duff to "seed" the soil around your new or problem plants with the beneficial and symbiotic fungi from where the same plants are thriving. Plants can usually live without their fungal and bacterial partners, but they won't thrive the same way they do when they have all the help they can get.

One can hope that the right yeast spore for making champagne will fall into the vat, but like professional bakers and winemakers, many growers use a culture that they trust to work and give them the results they expect.

There are a number of good sources for BSOs, including fungal mixes for special purposes such as growing conifers and specialized bacteria for making compost. We would recommend trying a few different types to decide which works best for you.

Nitrogen N
(also see *Estimating Nitrogen Release ENR* in appendix)

Many soil laboratories omit Nitrogen from the soil report; the soil report we are using for the examples in this book does not include N. The reason for this is that N levels are very unstable; Nitrogen is constantly leaking off into the air, leaching downwards out of reach of the roots, or simply becoming unavailable due to the soil temperature being too cold for biological action to release N from the organic matter reserves of the soil. Any test for Nitrogen provides only a snapshot of what is available at the time and at the temperature the test is done. In some cases, Nitrogen levels are simply estimated based on the soil's organic matter content. If the soil's organic matter content is 4% or above, there is likely to be a good amount of Nitrogen potentially available. Humus in the soil generally has a 10:1 Carbon to Nitrogen ratio.

Nitrogen is found in the soil in two forms, ammonia NH_4+ and nitrate NO_3-. NH_4+ is a cation base and can be held on the negatively charged soil colloids, humus, and clay. Nitrate NO_3- is highly water soluble and is more likely to be leached away; in addition microbial action is constantly changing NH_4 into NO_3. Nitrogen is a component of all proteins and amino acids. When living things die, their protein breaks down and ammoniacal NH_4 is released. Nitrate Nitrogen is associated with rapid growth while ammonia Nitrogen is associated with flowering, fruiting, and plant maturity.

Having an optimum level of **Sulfur** is important to conserve N in the soil. In the absence of adequate Sulfur, much of the ammonia N generated from the breakdown of organic matter in the soil or in compost will off-gas to the air and be lost. At the same time much of the Carbon will be lost as it off-gasses as CO_2. When optimum levels of S are present, a greater portion of N and C will be incorporated into stable humus and remain in the soil. The recommended Sulfur amount for the Ideal Soil method is ½ of Ideal Potassium, with a minimum of 50 ppm S for any productive agricultural soil.

The only way to get an accurate N test that will tell what is available in the soil right now is to express mail a sample to the lab, or to use a home testing system such as the LaMotte soil test, which is designed to measure only those elements immediately available in the soil, not the amount of reserves.

Nitrogen requirements vary greatly due to climate and crop, and should be addressed by individual crop needs. Some crops such as corn (maize) and alliums (onions, leeks, garlic) benefit greatly from an N boost or two during the growing season; for other crops this would only cause rank growth and delay flowering and maturity. Most crops do well with around 40 to 50 ppm available N, split evenly between nitrate and ammonia forms. (See Carey Reams' Ideal Soil recipe in the appendix.)

The only naturally occurring mineral source of Nitrogen is **Chilean nitrate of soda**, which contains around 16% N. It is a great source of natural N, especially in cold soils, but should not be used for more than 25% of Nitrogen needs due to its high level of Sodium. Chilean nitrate may not be allowed under some organic certification rules. If your farm is Certified Organic, check with your certifying agency before using Chilean nitrate.

The Ethics of Nitrogen in Organic Agriculture

Here are the N fertility inputs allowed under the USDA NOP organic rules:

Blood meal
Fish meal
Bone meal
Feather meal
Animal manures
Chilean nitrate of soda
and:
Oilseed meals: Soybean, flax seed, rapeseed, and cottonseed meals that are mostly GMO and have been processed with petroleum-derived hexane solvent to extract the oils. The meal left after extraction is used for animal feed as well as

"organic" agriculture. This seed meal contains up to 0.5% hexane residues, enough to kill baby pigs. Hexane is an extremely cheap byproduct of gasoline refining; Overton SV and JJ Manura (1997) found higher than expected levels of pentane, hexane, heptane, octane and benzene derivatives in all 6 hexane extracted cooking oil samples tested. If those solvents are in the vegetable oils, they are in the seed meals as well.

A moderate-sized 100,000 bushel per day soybean oil extracting facility can lose 6,000 pounds of hexane per day to the environment (atmospheric leaks from distillation, decanting, open vessels, and the meal).
http://www.karlloren.com/Diabetes/p47.htm

Only one USDA National Organic Program allowed source, Chile nitrate of soda, does not come from either industrial agriculture, industrial trawling of ocean fish, or confined animal feeding operations. Chile nitrate of soda needs to be imported from South America and is limited in usefulness because of its high Sodium content.

Does this make sense, that the only N sources allowed in what is supposed to be ethical and healthy agriculture come from GMO, chemically farmed, herbicide-sprayed seeds contaminated with hexane and other petroleum solvents, from giant factory ships sweeping up all sea life, or from grim, unnatural animal factories?

On the other hand, ammonium sulfate, urea and other "synthetic" Nitrogen fertilizers are readily available N sources made from atmospheric Nitrogen. They can be made anywhere rather than needing long distance transportation. No animals are mistreated in their manufacture. No oceans are seined by giant factory ships. They are not GMO, not chemically contaminated, and contain only pure forms of the desired plant nutrients.

Which Nitrogen source is cleaner, healthier, and more ethical to use?

Making "Synthetic" Nitrogen

The main process used for extracting N from the air, the Haber-Bosch, produces ammonia.

"The Haber process, also called the Haber-Bosch process, is the industrial implementation of the reaction of nitrogen gas and hydrogen gas. It is the main industrial route to ammonia:

$$N_2 + 3\,H_2 \rightarrow 2\,NH_3$$

Nitrogen is a critical limiting mineral nutrient in plant growth. Carbon and oxygen are also critical, but are easily obtained by plants from soil and air. Even though air is 78% nitrogen, atmospheric nitrogen is nutritionally unavailable because nitrogen molecules are held together by strong triple bonds. Nitrogen must be

'fixed', i.e. converted into some bioavailable form, through natural or man-made processes. It was not until the early 20th century that Fritz Haber developed the first practical process to convert atmospheric nitrogen to ammonia, which is nutritionally available."
http://en.wikipedia.org/wiki/Haber_process

The source of the N_2 is the atmosphere; the Hydrogen needed, as well as the source of heat, has traditionally come from methane, i.e. natural gas. Under high temperature and extremely high pressure (200+ atmospheres) the gaseous mixture of Nitrogen and Hydrogen is repeatedly passed over a bed of metallic Iron; the iron works as a catalyst to chemically bond the N and H, forming NH_3, ammonia.

Surely everyone has heard of "petroleum based fertilizers"? In reality, the Haber-Bosch process is the only industrial fertilizer process that has any "petroleum" connection, and that's a stretch, calling methane from natural gas petroleum. The refining processes for Phosphorus and Potassium fertilizers use no hydrocarbons.

Does "synthetic" N kill soil life? It can, when misused and overused. Probably the worst example is 'knifing' anhydrous ammonia into the root zone of maize/corn. While that can give a quick boost in growth and yield, the high dose of ammonia kills off most soil life, while burning up humus at the same time. Arden Andersen (*Science in Agriculture*) writes that injections of anhydrous ammonia were used in WWII in the Pacific to quickly turn jungle soils into hard-packed landing strips for aircraft.

On the other hand, adding a small amount of Ammonium sulfate to compost piles or to the soil can greatly accelerate microbial action in breaking down organic matter into stable humus. Both nitrate N and ammonia N occur naturally in all agricultural soils. We are aware of no evidence that there is any chemical difference between naturally occurring and industrially produced ammonia and nitrate.

A New Ecologically Sound Process for "Fixing" Atmospheric N

30 January 2013
Researchers from the University of Strathclyde and the University of St. Andrews have demonstrated that ammonia can be synthesized directly from air (instead of N2) and H2O (instead of H2) under a mild condition (room temperature, one atmosphere) with supplied electricity which can be obtained from renewable resources such as solar, wind or marine.....their process could also reduce the pressure on renewable energy storage, they note.

Globally 131 million tons of ammonia were produced in 2010. The dominant ammonia production process is the Haber-Bosch process invented in 1904 which requires high temperature (~500°C) and high pressure (150–300 bar), in addition to efficient catalysts. Natural gas or coal is used as the energy source of the ammonia industry. 1.87 tons of CO2 is released per ton of ammonia produced. In the Haber-Bosch process, the presence of ppm level oxygen may poison the commonly used Fe-based catalysts. In industry, extensive purification of N2 and H2 is needed and this remarkably increases the overall cost of the process. Therefore researchers have been seeking a simpler way for synthesis of ammonia from nitrogen separated from air.

In this study, the researchers first fabricated an electrochemical cell for ammonia synthesis. H2 (or water) and N2 (or air) were passed through room temperature water first then filled into the chambers of the cell.

A maximum ammonia production rate.....was achieved when a voltage of 1.6 V was applied. "In conclusion, for the first time, this experiment clearly indicates that ammonia can be directly synthesised from air and water at room temperature and one atmosphere."
http://www.greencarcongress.com/2013/01/ammonia-20130130.html

Potentially, this can provide an alternative route for the mass production of the basic chemical ammonia under mild conditions. Presently existing wind power generators that today have to be powered down when there is no need for their electricity could instead be put to work synthesizing ammonia fertilizers.

Rong Lan, John T. S. Irvine & Shanwen Tao (2013) Synthesis of ammonia directly from air and water at ambient temperature and pressure. Scientific Reports 3, Article number: 1145 doi: 10.1038/srep01145
Original paper at http://www.nature.com/srep/2013/130129/srep01145/full/srep01145.html

A printable periodic table of the elements in .PDF form may be found here:
http://www.webelements.com/nexus/sites/default/files/webelements_table_5sf_2012-06-07.pdf

Chapter 7
Converting from ppm to kilograms/hectare, pounds/acre or other weights and volumes

Since Chapter 3 we have been working with parts per million. In order to write a fertility Rx for our soil or growing media we need to convert ppm into the units per area or volume we wish the final Rx to be written in.
The table below shows the amount of minerals that we have calculated need to be added, converted to weight per unit area for both the metric and pounds/acre systems.

Mineral	**Amount Needed PPM**	**Kg/ha**	**Kg/100 m^2 (Kg/ha ÷ 100)**	**Lbs/Acre**	**Lbs/1000 ft^2 Lb/ac ÷ 43.56**
Calcium	300	600	6	600	13.77
Magnesium	48	96	0.96	96	2.20
Potassium	89	178	1.78	178	4.09
Sodium	13	26	0.26	26	0.60
Phosphorus	78	156	1.56	156	3.58
Sulfur	69	138	1.38	138	3.17
Boron	1.34	2.68	0.027	2.68	0.06
Iron	39	78	0.78	78	1.79
Manganese	34	68	0.68	68	1.56
Copper	7.13	14.28	0.143	14.28	0.33
Zinc	0	0	0	0	0

Recall that we are working with the upper 2 million kilos of 1 hectare of land, or the upper 2 million pounds of 1 acre. One millionth of that mass is 2 kg /ha or 2 lbs/acre.

Kg per Hectare to Kg per Are

One hectare is 10 000 meters^2 (100m x 100m). 1/100th of that is 100 m^2, an area of measure also known as an **are** (pronounced 'ar' or 'air' in English). **100 ares = 1 hectare.** This makes the **conversion from kg/hectare to kg/are** simple: just **move the decimal point 2 spaces to the left.**

Sulfur needed: 138 kg/ha = 1.38 kg/are. **(1 Are = 100 m^2 = 1076 sq ft)**

Lbs per Acre to Lbs per 1000 ft^2

One acre is 43 560 ft^2. There are 43.56 x 1000 ft^2 sections in an acre. **To convert lbs/acre into lbs/1000 ft^2**, we **divide lbs/acre by 43.56:**

Sulfur needed: 138 lbs / 43.56 = 3.168 lbs, rounded off to two decimals, 3.17 lbs/1000 ft^2.

Working With Volume: Cubic Meters and Cubic Yards

To amend the minerals in potting soil, container growing mixes, raised beds or compost piles, one needs to know the weight per cubic volume of the dry soil or growing media.

If one is working with **mineral soil** in the **pounds/acre system**, it's safe enough to assume that **an acre of soil 6 inches deep** will weigh **2 million pounds**.
2 000 000 lbs / 43 560 sq ft = 45.91 lbs per square foot. Two 6 inch deep 1 square foot sections would be 1 **cubic foot** which would weigh about 92 lbs, and there would be
43 560 / 2 = **21 780 cubic feet in the top 6" of an acre of soil.**

A cubic yard is 3 ft x 3 ft x 3 ft = 27 cubic feet.

21 780 ft^3 / 27 = **806.67 cubic yards in the top 6" of an acre of soil**; close enough to 800 cubic yards.

For *mineral* soils, if we know **how many lbs**/acre of a mineral or amendment we need, we can **divide lbs/acre by 800** to find out **how many lbs/cubic yard** we need:

We need 600 lbs/acre of Calcium.
600 lbs per acre / 800 yds^3 per acre = 0.75 lbs Ca / $yard^3$.

For a **mineral soil** in the **kilogram/hectare system**, we are working with 10 000 square meters of soil ~15 cm deep that weighs 2 million kg. **100 centimeters = 1 meter.** 15 cm is 0.15 meters.. To calculate the number of cubic meters in 1 hectare to a depth of 15 cm we multiply
10 000 m x 0.15 = **1500 cubic meters per hectare** to a depth of 0.15 m.

If we know **how many kg** of an element or amendment is **needed per hectare, in a mineral soil,** we can **divide by 1500** to see how many **kg** the soil needs **per cubic meter:**

We need 600 kg/ha of Calcium
600 kg / 1500 = 0.4 kg per cubic meter

The **average weight per cubic meter** of a **mineral soil** will be 2 000 000kg / 1500 m^3 = **1333 kg/m^3**.

Working With Lightweight Mixes and High Organic Matter Soils

The figures above are close enough for heavy mineral soils, but naturally high-organic matter soils such as peat soils or virgin forest soils may weigh much less than 92 lbs per cubic foot or 1333 kg per cubic meter. Lightweight potting media may weigh as little as 10 lbs/ft^3 (270 lbs/yd^3) or 150 kg/m^3.

If you are working with a light weight soil, potting media, or compost/organic matter it is best to measure and weigh an oven-dry sample. Pack a 1 gallon or a 4 or 5 litre container with the damp sample (as it will be when in use), oven dry at 250* F (120*C) until very dry, then weigh the oven dry media and record the weight in grams.

Calculating the Amount Needed Per Cubic Yard: Lightweight Soils

There are 7.48 (~7.5) US Gallons (128 fluid oz) per cubic foot and 201.96 (~200) gallons per cubic yard. If a gallon of soil weighs 2 lbs, a cubic foot will weigh 7.5 gallons x 2 lbs = 15 lbs. At 200 US gallons per yard^3, 2 lbs x 200 gallons = 400 lbs per cubic yard.

Once we know the weight per cubic yard we can calculate how many pounds or fraction of a pound equals 1 ppm. The cubic yard of growing media above weighs 400 lbs. One-millionth of that: 400 lbs / 1 000 000 = 0.0004 lbs; 4 ten-thousandths of a pound. A good digital scale can weigh that amount, but it's awkward. We could choose to work with ounces and decimal fractions of an ounce:

1 lb = 16 oz, so 0.0004 lb x 16 = 0.0064 oz

but it's usually simpler and easier to convert the pound weight to kilograms and grams for small amounts of soil media or amendments

Our cubic yard above weighed 400 lbs. **1 kg = 2.20 lbs.**
Divide 400 lbs by 2.2 kg/lb = 182 kg per cubic yard.

A kilogram is 1000 grams. There are **1000 milligrams in a gram**.
1000 x 1000 = 1 million. **A milligram is 1 ppm of a kilogram.**

Our worksheet says we need 300 ppm Calcium. For **each kilogram** of weight, **we need** 300 ppm or 300 milligrams **(0.3 grams) of Calcium**.
Multiply 182 kg x 0.3 grams/kg = 54.6 grams of Calcium needed per cubic yard.

Calculating the Amount Needed Per Cubic Meter

There are 1000 liters per cubic meter. The 5 liters we talked about weighing above are equal to five thousandths or 0.005 of 1 cubic meter. If 5 liters of soil weighs 1 kg, the weight per cubic meter would be 1 kg / 0.005 = 200 kg.

200 kg x 0.3 gram/kg = 60 grams of Calcium needed per m^3.

Calculating Weight Per Depth of Light-Weight Soils

Suppose we were working with an acre or hectare of loose-textured high-organic matter soil that, rather than weighing 92 lbs/ft^3 or 1333 kg/m^3, **weighed only 60 lbs/ft^3 or 870 kg/m^3?**

Obviously a 6 inch or 15 cm depth of this soil would not weigh 2 million lbs/acre or 2 million kg/hectare, but for the sake of simplicity, we choose to work with 2 million lbs or kg. This poses two questions: **"How deep is 2 million lbs or kg of this soil?"** and **"How much does the top 6" or 15 cm actually weigh?"**

The answer to the second question is easy enough. We figured out above that there were 1500 cubic meters per hectare to a depth of 15 cm. Multiply 870 kg/m^3 x 1500m^3 = 1 305 000 kg/ha to a depth of 15 cm.

We also calculated above that an acre 6" deep is 21 780 ft^3;
60 lbs/ft^3 x 21 780 ft^3 = 1 306 800 lbs/ac to a depth of 6" (the difference between the numbers for lbs/ac and kg/ha here are due to rounding off earlier).

We see that the top 6" or 15 cm of this soil weighs about 700 000 lbs/ac or kg/ha (35%) less than our "standard" of 2 million. If we decided to amend only the top 6" or 15 cm of the soil, we would take our amendment weight needed for 2 million weight units and multiply by 65%. For example, 600# Ca x 0.65 = 390# Ca needed to bring the top 15cm or 6" of this light-weight soil into balance.

The other question was **"How deep is 2 million lbs/ac or kg/ha of this soil?"**

If 15 cm or 6" of the soil weighs 65% as much as our standard soil, then 2 million lbs or kgs of this soil would cover a depth equal to the standard depth divided by 0.65:
15 cm / 0.65 = 23 cm deep, or 6" / 0.65 = 9.23" deep. If we add the required 600 lbs/ac or kg/ha Calcium, that will be enough to balance the Ca in the top 23cm or 9.23 inches of this soil.

A printable periodic table of the elements in .PDF form may be found here:
http://www.webelements.com/nexus/sites/default/files/webelements_table_5sf_2012-06-07.pdf

Chapter 8

Writing the Soil Rx

Time to put it all together: We have calculated what to add to this soil to bring it into line with the Ideal Soil; now we need to decide which amendments will do the job for the least amount of effort and expense.

There is no truly simple method of deciding what source to use for many of these minerals. The amendments shown in the Typical Analysis chart on the next page, all of which are allowed under the USDA NOP (National Organic Program), are mostly mixtures. One needs to look at various combinations and possibilities while keeping availability and cost in mind. What is locally available at a fair price should be used whenever possible.

Here are our worksheet calculations from chapters 3-6 converted from ppm to kg/ha or lbs/acre. **The hash mark # indicates either lbs/acre or kg/ha.**

Mineral	Amount Needed PPM	Kg/Ha or Lbs/Acre
Calcium	300	600#
Magnesium	48	96#
Potassium	89	178#
Sodium	13	26#
Phosphorus	78	156#
Sulfur	69	138#
Boron	1.34	2.68#
Iron	39	78#
Manganese	34	68#
Copper	7.13	14.28#
Zinc	0	0

On the next page is a table showing the typical mineral content of common soil mineral amendments allowed under the USDA NOP (National Organic Program) rules, **Subpart G—Administrative**
The National List of Allowed and Prohibited Substances § 205.600--601

Typical Mineral Content of USDA Organic Fertilizer Ingredients (%)

Animal Source	N	P as P_2O_5	K as K_2O	S	Ca	Mg	Fe	Tr
Fish Bone Meal	4	20		0.6	19	0.3		Tr
Fish Meal	10	4.5		0.6	2.3	0.3		Tr
Crab Shell	3	3.25	0.3	0.2	23	0.3		Tr
Blood Meal	13	1						
Feather Meal	12	0.1	0.4	0.4	0.6	0.6		
Bone Meal	3	15			20	0.4		
Mineral Amendments and Kelp								
	N	P as P_2O_5	K as K_2O	S	Ca	Mg	Fe	Tr
Ag Lime					32-40	1-5		
Dolomite Lime					22	13		
Gypsum*				16	22			
Oyster shell					36	0.3		
Epsom salt**				14		10		
Potash sulfate**			51	17.5				
TN brown phos		3 (23% total)			40			Tr
Calphos		3 (20% total)			20			Tr
K Mag*			22	22		11		
Greensand		1	7		1.3	2.2	9	Tr
Kelp Meal	1	0.7	3	2	2	0.7		Tr

Tr = Good source of micro (trace) minerals

Purified Source	Sulfur S	Boron B	Iron Fe	Mang. Mn	Copper Cu	Zinc Zn
Ag Sulfur	**90**					
Borax**		**9**				
Solubor™**		**20.5**				
Fe sulfate $1H_2O$	**18**		**30**			
Fe sulfate $7H_20$**	**11.5**		**20**			
Mn sulfate $1H_2O$*	**19**			**32**		
Cu sulfate $5H_2O$**	**12.5**				**25**	
Zinc sulfate $1H_2O$	**17**					**35**
Zinc sulfate $7H_20$**	**11**					**22**

****Highly soluble in H_2O *Varies in solubility in H_2O**

For the rest of this chapter, the amounts we calculate will be interchangeable between kg/ha and lbs/acre. **The hash mark # indicates either lbs/acre or kg/ha.**

We will begin with the major cations Ca, Mg, and K.

Calcium Ca and Magnesium Mg

Calcium Ca++ ppm	Desired Found Deficit	1550 1250 -300	Add 300 ppm **(600 #)** Calcium Ca
Ca Base Saturation 60-70 %		55.0%	
Magnesium Mg++ ppm	Desired Found Deficit	164 116 -48	Add 48 ppm **(96 #)** Magnesium Mg
Mg Base Saturation 10-20 %		8.5%	

Below is the Ca and Mg section from our worksheet.

Magnesium can be difficult to balance when writing a Certified Organic soil Rx as there are no USDA NOP approved sources that are high in Mg. Here are our choices:

Dolomite lime: Ca 22%, Mg 13%
Epsom Salts (Magnesium sulfate): S 13%, Mg 10%
K-Mag (Sulfate of Potash Magnesia): 22% K_2O, 22% S, 11% Mg.

Note that all three approved sources are running 10-13% Mg. That means we are going to need 75-10 # of any of them to end up with 10# of Mg. As this soil requires 96 # of Mg, we are going to need up to 960# depending on which source we choose.
If we needed only Magnesium, no Ca, S, or K, the USDA NOP rules would allow the use of naturally occurring mined Magnesium carbonate $MgCO_3$, magnesite, which contains up to 20%Mg, but this is not commonly available. If government organic certification is not an issue, Magnesium oxide MgO, at around 50%Mg, is pure, safe, inexpensive, readily available, and is rapidly assimilated into the soil.
Here are some 2013 retail prices from the USA:
K-Mag 24 kg (50 lbs): $25
Epsom Salts 24 kg: $25
Dolomite Lime 24 kg: $8

At first glance it looks like Dolomite is our best choice if we are amending a large field. If we are amending a 1000 ft^2 or 100 $meter^2$ garden, the price might not be a major consideration. Other considerations enter into the decision. Dolomite lime is about 22% Calcium, so for every 13# of Mg we are adding 22# of Ca. If our worksheet told us that we didn't need any Ca, we would have to use one of the other Mg sources. However, both of the other choices contain more Sulfur than they do Magnesium and the K-Mag contains almost twice as much Potassium as Magnesium. This soil does need Potassium, and Sulfur, and as a matter of fact we need almost twice as much K as Mg: 178# K and 96# Mg. We only need 138# of Sulfur, but a little extra Sulfur won't hurt.

It appears that K-Mag might be a good choice, even though it is more expensive than dolomite lime, because we are getting the Sulfur and the Potassium we need at the same time. How much K-Mag would we need? We will take the amount of Mg we need and divide it by the % content of Mg in K-Mag:

96# Mg / 0.11 = 873# of K-Mag.

That's a lot of money at $1 per kg, almost $900 per hectare. For the 8.73 kg needed in a 100 sq meter garden it would only cost 1/100th that much, around $9, and would take care of Mg, K, and S all at once. For a small garden it sounds like a good deal; for a large area, no, unless the grower has access to KMag at a much lower price. Dolomite lime can be ordered in bulk, which would bring the price down considerably. **As we also need to add Calcium, it seems best to use Dolomite.** How much 13% Mg dolomite do we need?

96# Mg / 0.13 = 738# dolomite lime is needed (dolomite lime is available in the USA in bulk for less than $50/ton)

At the same time we are getting **22% Calcium** in the dolomite lime:

738# x 0.22 = 162# Ca

We need 600# of Calcium so that leaves us still 438# short of Ca.

The least expensive form of Calcium is agricultural "sweet" lime, Calcium carbonate (i.e. high-Calcium ag lime, ground-up limestone), which according to the Typical Analysis chart is 32% to 40%% Ca. We will assume the lime we will use is 39% Ca:

438# / 0.39 = 1123# high-Calcium agricultural sweet lime is needed.

Calcium Ca++ ppm	Desired Found Deficit	1550 1250 -300	Add **1123#** high-Calcium ag lime 39%Ca
Ca Base Saturation 60-70 %		55.0%	
Magnesium Mg++ ppm	Desired Found Deficit	164 116 -48	Add **738#** Dolomite lime 13%Mg 22%Ca
Mg Base Saturation 10-20 %		8.5%	

Potassium K and Sulfur S

From the worksheet:

Potassium K+ Ppm	Desired Found Deficit	178 89 -89	Add 89 ppm **(178#)** Potassium K
K Base Saturation 2-5 %		2.0%	

The most readily available and least expensive, USDA organic approved Potassium source is Potassium sulfate (usually known in the fertilizer trade as **Sulfate of Potash** or just SOP) which **is 51% potash (K2O), and 17.5% Sulfur**. The fertilizer label in the USA would read **0-0-51 17.5S**. Earlier we determined that potash was 83% elemental K, so the actual amount of K in Potassium sulfate is 83% of 51%:

0.51 x 0.83 = 0.423 or 42.3% actual K

We need 178# of K:

178 / 0.423 = 421#
We need to add 421# of Sulfate of Potash 0-0-50 to this soil.

Potassium K+ ppm	Desired Found Deficit	178 89 -89	Add **421**# Sulfate of Potash 0-0-51
K Base Saturation 2-5 %		2.0%	

According to our table of Typical Mineral Content, Sulfate of Potash is 17.5% S, so the amount of Sulfur we are getting in this K2SO4 is

421 x 0.175 = **74# Sulfur**

Our worksheet calls for adding 138# of S, so we are still 63# short. This could be remedied by adding 70# of 90% ag Sulfur, (63# / .9 = 70#) but let's hold off on that. We still have 78# of Iron, 68# of Manganese, and 14.28# of Copper to calculate into the prescription, all of which will be added in sulfate form.

Sodium Na

This soil needs **26# of Sodium**. Sea salt or mined salt deposits from ancient sea beds are good trace mineral sources. Sea salt contains at least 80 different minerals.

Sea salt is around 40% Sodium, depending on the moisture content.

26# / 0.40 Na = 65#

We need to add 65# sea salt or mineral salt.

Sodium Na+ ppm	Desired Found Deficit	39 26 -13	Add **65#** sea salt or **96#** Chilean Nitrate of Soda
Na Base Saturation 1-5 %		1.0%	

Another source of Sodium: Chilean Nitrate of Soda
Chilean nitrate of soda is a naturally occurring Sodium nitrate mineral deposit that is mined in the high, dry deserts of Chile and Peru. It contains **16% soluble Nitrogen** as nitrate NO_3 and **27% Sodium** by weight. Mined Sodium nitrate is allowed for use under USDA NOP rules as long as it does not supply more than 20% of the soil's annual nitrogen needs. If we used nitrate of soda to supply our needed Sodium, we would need

26 / 0.27 Na = **96# Chilean nitrate**, which is 16%N and would supply 96 x 0.16 = 15# of readily available Nitrogen along with the 26# of Sodium.

Phosphorus P

The soil worksheet says that we need to add 78ppm or **156# elemental Phosphorus**. In **Chapter 4** on Phosphorus, Sulfur, and Chlorine we determined that **phosphate, P_2O_5, is 44% Phosphorus** by weight.

Florida clay phosphate is 20% P_2O_5; 44% of 20% is:

0.20 x 0.44 = 0.088 or **9% actual P**

Tennessee brown phosphate is 23% P2O5; 44% of 30% is:

0.23 x 0.44 = 0.10 or **10% actual P**

The choice we make will depend on the type of soil we are working with, our proximity to the source, and the price at which it is available. As of 2013, the "rock" phosphate **most commonly available in the US is Calphos** brand colloidal clay phosphate from Florida, which we determined was 9% Phosphorus.

156 / 0.09 = **1733# of Calphos soft rock phosphate needed**.

1733 lbs/ac or kg/ha is a lot of phosphate rock.. Is it too much to add all at once? No, because the readily available P2O5 is only 3%, so we are really applying only 60# of "available" P2O5.

It is **generally safe to add 2200 kg/ha or 2200 lbs/ac** (22 kg/100m^2 or 50lbs/1000ft^2) of any **phosphate rock to any soil in a single application**. Up to twice that much may safely be added at once if the phosphate rock is a non-reactive type like Calphos. If one needs to add a large amount of reactive rock phosphate it should be split into two or more applications a few months apart, and the soil's pH and Ca levels monitored, as reactive rock phosphates frequently have a net acid reaction in soils.

Calcium Content of Phosphate Rock
The question arises: Should the amount of **Calcium in the phosphate** be counted as contributing to **Exchangeable Calcium?** If it is a reactive phosphate, all of the Ca will be chemically bonded with P or other anions. **Ca will be released only when the chemical bond is broken and the P is made available; it is not available in the soil solution**, not a free cation, and not available to fill a negative exchange site. The same applies to phosphate sources like fish bones or crab shells: the P and Ca are chemically bonded and the Ca is not readily available. Non-reactive phosphates like Calphos may have some free Calcium but generally not enough to remedy a Calcium deficiency. **As a rule, do not count the Calcium in any phosphate source when calculating the amount of Ca being added to a soil.**

Phosphorus P ppm	100	Add **1733#** Calphos soft rock phosphate

We will add 1733 kg/ha (or 1733 lbs/acre) of Calphos soft rock phosphate

So far we have determined the amendments needed to bring us to the Ideal Soil balance for the minerals P, Ca, Mg, K, and Na. We have provided part of the S, but still need another 63 kg/ha or lbs/acre of Sulfur to reach our ideal level where S = ½ of Ideal K. The extra S needed will come from the sulfate forms of Iron, Manganese, and Copper we will be adding.

Next we need to calculate the secondary elements Boron, Iron, Manganese, Copper, and Zinc.

Boron B

There are several Boron compounds produced especially for agriculture, ranging up to 20% B by weight. The easiest form of Boron to find in the USA, however, is regular borax powder (sodium borate, **20 Mule Team Borax**) sold in the soap and detergent section of grocery stores in the USA. It averages about **9% B** by weight.

The worksheet calls for adding 1.34 ppm, which is 2.68 kg/ha or lbs/acre. Using regular mined borax, Sodium borate 9%B, we will need 2.68 / 0.09 = 29.77# borax (round up to 30#). **We need to add 30# borax 9% B.**

Solubility of Iron, Manganese, Copper and Zinc Amendments

Most Fe, Mn, Cu, and Zn soil amendments are applied as sulfates. The purified metals have been reacted with sulfuric acid and water to form more or less soluble compounds. As a rule, the more H_2O molecules attached to the sulfate, the higher the solubility. Zinc sulfate monohydrate, $ZnSO_4$ $1H_2O$ is somewhat soluble in water, Zinc sulfate heptahydrate, $ZnSO_4$ $7H_20$ is highly soluble. Which one we choose will depend on whether we want slow release Zinc or highly soluble Zinc, whether or not we will be able to till the amendment into the soil, and whether the soil needs more Sulfur or already has too much. The oxide forms of the metals are not soluble in water, but are a good choice for soils that are naturally high in sulfates, e.g. gypsum soils.

Iron Fe

The worksheet calls for 39ppm Fe = 78# Fe

Standard feed grade **ferrous sulfate monohydrate** FeSO4 1H2O is **30% Fe and 18% Sulfur.** It is only sparingly soluble in water. **Ferrous sulfate heptahydrate**, $FeSO_4$ $7H_2O$ (also known as "copperas" due to the green color of its crystals) is highly water soluble but only contains **20%Fe and 11.5%S** by weight. A third alternative for soils high in sulfates where one does not wish to add more S, is **Iron oxide** Fe_3O_4, which is around **70% Iron**.

We will choose to use the water-soluble heptahydrate, copperas, which is 20%Fe.

78# / 0.20 = 390# $FeSO_4$ $7H_20$

Iron Fe	50	Add **390#** Ferrous sulfate 20%Fe

We need to add 390 kg/ha or lbs/acre Iron sulfate 20% Fe

In addition, this iron sulfate contains 11.5% Sulfur

390 x 0.115 = 44.85 rounded to 45 lbs or kg Sulfur. After calculating the Potassium sulfate we needed another 63 units of S. 63 – 45 = 18. Now we only need another 18 units, which will be taken care of with the addition of Manganese sulfate and Copper sulfate. **We will have more than enough Sulfur when all of the amendments are added.**

Boron B	0.21	Add **30#** Borax 9% B

Note on Iron: Other common mineral amendments such as greensand and many rock dust powders contain significant percentages of Iron. If you have easy access to these in quantity you may be able to meet the soil's iron requirements with them instead of buying the more expensive purified Iron sulfate.

Manganese Mn

The worksheet says 34ppm or **68# of Manganese** is needed.

Manganese Mn	11	Add **212#** Manganese sulfate 32%Mn

The **Manganese sulfate monohydrate** commonly available is 32% Mn and 19% S

68# / 0.32 = **212# of $MnSO_4$ $1H_2O$ is needed**.

In addition we will be gaining 19% of this weight in Sulfur

212 x 0.19 = 40 kg/ha or lbs/acre of Sulfur. We will have plenty of Sulfur. A little extra won't hurt.

Copper Cu

We need 7ppm or **14#** of Copper
Our Copper sulfate amendment contains 25% Cu

Copper Cu	1.07	Add **56#** Copper sulfate 25%Cu

14 / 0.25 = 56#

We need to add 56# Copper sulfate 25% Cu

Zinc Zn

Our Ideal P=K number is 178ppm. Zn should be $1/10^{th}$ of that, or 17.8ppm. The lab test found 16.4ppm Zn, close enough. **We do not need to add any Zinc**

Our calculations for the soil mineral prescription are finished. Hearty congratulations to the reader who has made it this far.

The Final Product: The completed soil prescription

On the next pages you will find the completed soil report and recommendations in the format used when writing soil prescriptions at SoilMinerals.com.

You will note that the trace mineral sources, beneficial organisms, and humic acid sources mentioned above have been added as "optional but recommended".

Instructions for applying the amendments are always a good idea and have been included.

The format that we use for the soil report and worksheet is a slightly modified form of the one developed by Wm Albrecht, Louis Bromfield, and their associates in the 1940s. You will find a blank soil report worksheet in the appendix.

The completed soil prescription

Element		Results	Comments
Cation Exchange Capacity CEC meq		11.4	
pH of Soil Sample		5.58	
Organic Matter %		5.6%	
Primary Anions			**All Amendments are kg/ha or lbs/acre #**
Sulfur S (parts per million ppm)		20	Will be supplied by sulfate amendments
Phosphorus P ppm		100	Add **1733#** Calphos soft rock phosphate
Primary Cations			
Calcium Ca++ ppm	Desired Found Deficit	1550 1250 -300	Add **1123#** high-Calcium ag lime 39%Ca 1254 or
Ca Base Saturation 60-70 %		55.0%	54.8%
Magnesium Mg++ ppm	Desired Found Deficit	164 116 -48	Add **738#** Dolomite lime 13%Mg 22%Ca
Mg Base Saturation 10-20 %		8.5%	
Potassium K+ ppm	Desired Found Deficit	178 89 -89	Add **421#** Sulfate of Potash 0-0-51
K Base Saturation 2-5 %		2.0%	
Sodium Na+ ppm	Desired Found Deficit	39 26 -13	Add **65#** sea salt or **96#** Chilean Nitrate of Soda
Na Base Saturation 1-5 %		1.0%	
Other Bases		6.2%	
H+ Exch Hydrogen 10-15%		27.0%	
Secondary Elements ppm			
Boron B		0.21	Add **30#** Borax 9% B
Iron Fe		50	Add **390#** Ferrous sulfate 20%Fe
Manganese Mn		11	Add **212#** Manganese sulfate 32%Mn
Copper Cu		1.07	Add **56#** Copper sulfate 25%Cu
Zinc Zn		16.4	OK
Aluminum		1841	Normal

Soil Report and Comments

Client: **Location:** **Soil Test:** **Date:**

Notes: Soil amendment recommendations based on Ideal Soil chart and worksheet in The Ideal Soil v2.0

USDA Organic Approved Nutrients Recommended, kg/ha or lbs/acre:

Calphos soft rock phosphate: 1733#

High Calcium ag lime 39%Ca: 1123#

Dolomite lime 13%Mg: 738#

Sulfate of potash 0-0-51: 421#

Sea salt: 65# (or 96# Sodium nitrate)

Borax 9%B: 30#

Ferrous sulfate 20%Fe (heptahydrate, copperas): 390#

Manganese sulfate 32%Mn (monohydrate): 212#

Copper sulfate 25%Cu: 56#

Optional:

Azomite trace minerals: 400#

Humate ore: 400#

Kelp meal: 400#

Feather meal: 800#
(or other Nitrogen source to supply ~100# N)

All of the above may be mixed together for even application. If possible, they should be well blended into the top 4" to 6" (10 to 15cm) of soil and allowed to settle in for a couple of weeks before seeds are planted. Sturdy transplants can go in at any time.

How Close is "Close Enough"?

How important is it to apply the exact amounts we have calculated? Not all that important. Soil is not a homogenous material. It can vary significantly within the same garden bed, or even within a few inches. In addition, even the best soil tests are not 100% accurate. If the amounts applied are within 10 or 15% of the amounts calculated, all will be well. The soil Rx on the previous pages calls for 1733# of Calphos and 421# of sulfate of potash. One could apply 1700# or 1800# of Calphos, or 400# or 450# of sulfate of potash without any problems. In practice, we usually round off the amounts recommended to an "even" number, for example:

The 56# of Copper sulfate would be rounded off to 55# or 50#, when working with acres or hectares.

When working with 100 m^2: 56 kg / 100 = 0.56 kg, which could be rounded to 0.55 kg or 0.50 kg.

When working with 1000 ft^2, 56 lbs / 43.56 = 1.286 lbs, which can be rounded to 1.30 lbs or 1.25 lbs.

This concludes the main "how to" portion of The Ideal Soil: A Handbook for the New Agriculture. If it all seems a bit much at this point, be assured that it will become easier as you work with the system. It is not simple or simplistic, just as Nature is not simple, nor is nutrition, but after you become familiar with the mineral amendments that are available and work with a few soil tests you will learn to juggle the various possibilities much more easily.

Also be assured that this system works and works very well. After you have balanced the minerals in a soil one time, further additions will be minor. Expect good results the first year, and increasingly better results for years afterward. Know that if this system is followed you will not have to worry about nutrient deficiencies in your crops or your food, and you will have the highest quality produce that can be grown. When word gets out, expect the world to beat a path to your door.

If it is not economically feasible to balance the minerals on your whole garden, pasture, or crop acreage, do part of it, what you can afford. The most important ones to get right, Calcium and Magnesium, are also the least expensive, so do them first.

Congratulations on becoming a pioneer of the New Agriculture.

Conversions:

1 acre = 43 560 ft^2

To convert lbs/acre to lbs/1000 ft^2, divide by 43.56 or 44

1 acre = 0,405 hectare

1 hectare = 2.47 acres

1 hectare = 10 000 $meters^2$

100 $meters^2$ = 1 Are (air) = 1076 ft^2

The top 6 to 7 inches of 1 acre of average soil is assumed to weigh 2 000 000 lbs

The top 15 to 17 cm of 1 hectare of average soil is assumed to weigh 2 000 000 kg

1 part per million (ppm) = 2 lbs/acre or 2 kg/hectare

1 ppm = 2 grams per 100 ft^2 or 2 grams per 10 m^2

1 ppm = 20 grams per 1000 ft^2 or 20 grams per 100 m^2

1 lb/acre = 1 gram per 100 ft^2

1 kg/ha = 1 gram per 10 m^2

Chapter 9
Calcareous and High pH Soils

Balancing soil minerals to the Ideal Soil ratios requires three sets of information:

1. An accurate assessment of the potentially available soil reserves of eleven elements: S, P, Ca, Mg, K, Na, B, Fe, Mn, Cu, and Zn.as well as soil pH

2. An accurate measurement or estimate of the soil's CEC, its functional Cation Exchange Capacity.

3. An accurate measurement of the ratio and amount of the base cations Ca, Mg, K, and Na presently held on the negative exchange sites in the soil. This is the Base Cation Saturation Ratio, Albrecht's BCSR.

Number1 is straightforward enough. There are several soil test methods or combinations of tests that can measure available or potentially available soil minerals. The Ideal Soil ratios were developed using the Mehlich 3 test, which gives a reasonably accurate assessment of all eleven elements in most soils below pH 7. Problems arise when the Mehlich 3 or any of the common multi-element soil tests are used to estimate CEC and exchangeable cations in soils above pH 7 and soils with undissolved limestone particles.

A soil's Cation Exchange Capacity is the sum total of the negative charges available to make an electrostatic bond, an ionic bond, with a positively charged ion. None of the common soil tests actually measure the soil's total negative charge. What they do instead is extract the base cations Ca, Mg, K, and Na (along with other elements) from the soil. The sum of the extracted base cations is used to estimate the cation exchange capacity of the soil. Obviously this estimate will only be valid if the amounts of Ca, Mg etc being used for the calculation have actually been extracted from negative exchange sites. In soils of pH 7.0 and below this will usually be the case. To understand why we need to review what the symbol pH stands for.

pH (potenz **H**ydrogen**) is the ratio and concentration of H+ Hydrogen ions to OH- hydroxyl ions in an H_2O (water) solution**. Another way of saying that is: pH is the proportion and amount of negative charges compared to positive charges in a water solution. At pH 7.0, + and - are equal and balanced. If there are more H+ ions than OH- ions, the pH is below 7 and is called acidic. If there are more OH- than H+ ions the pH is above 7 and is called alkaline or basic. Importantly, the solution will always strive for equilibrium, for all of the charges to be equalized. Excess OH- ions will be attracted to and react with a source of positive charge +, excess H+ ions will be attracted to and react with a source of negative charge -. This will continue until all unbalanced charges have been neutralized, or until there are no more readily available sources of + or – charges to react with.

The Technical Definition: pH is a logarithmic measure of hydrogen ion concentration, originally defined by Danish biochemist Søren Peter Lauritz Sørensen in 1909.
pH = -log[H+]
where log is a base-10 logarithm and [H+] is the concentration of hydrogen ions in moles per liter of solution. According to the Compact Oxford English Dictionary, the "p" stands for the German word for "power", potenz, so pH is an abbreviation for "power of hydrogen". In Sørensen's original paper, pH is written as P_H. The modern notation "pH" was first adopted in 1920 by W. M. Clark for typographical convenience.

How pH works in the soil

Whether a soil is acid or alkaline depends largely on rainfall and temperature. All rain and snow is acidic, below pH 7, if only because it has CO_2 gas dissolved in it. CO_2 gas dissolved in water, H_20, forms carbonic acid, H_2CO_3, which quickly dissociates to HCO_3- and H+. If the soil is alkaline, i.e. it has an excess of OH- ions, these will be neutralized by the free H+ ions from the carbonic acid in the rainfall. As the rain continues to fall on an alkaline soil, more and more OH- ions will combine with H+ to form stable H_20 and the pH (the ratio of OH- to H+ ions in solution) will drop. When there are no more OH- ions readily available for the H+ to combine with, the H+ ions from precipitation will begin to be in excess and will be drawn to and start reacting with carbonate rocks or any other alkaline material in the soil.

The reaction with Calcium carbonate limestone $CaCO_3$ in the soil goes like this:

$CaCO_3$(solid) + CO_2(gas) + H_2O(liquid) --> Ca++ (aqueous) + 2 HCO_3- (aqueous). (Aqueous = dissolved in water.)

The newly-soluble Ca++ can be attracted to and held on negative exchange sites on clay or humus, or it may be washed away to a lower soil horizon, depending on whether there are available negative exchange sites to adsorb it, and on rainfall and evaporation. In soils that have more rainfall than evaporation, eventually most of the alkaline rocks will have their base cations extracted into the soil solution by reacting with carbonic acid (or other acids in precipitation such as sulfuric acid H2SO4). If the soil has abundant negative exchange sites available, many of the soluble base cations will be held on those sites, keeping them from leaching away, and where they remain readily available as nutrients to plants and soil organisms. If the soil has a low CEC there will be little charge to hold the soluble cations and they will quickly be washed away and lost.

At the point where most of the alkaline rocks have had their base cations extracted, or where the amount of alkaline elements available balances the acidity of the precipitation, the soil reaches equilibrium of + to - charges and a pH of 7.

As long as precipitation exceeds evaporation the acidification and leaching process continues. The excess of H+ ions in the soil will increase and they will begin to displace the base cations Ca, Mg, K, and Na from the negative exchange sites. First the Calcium and Sodium are leached away, leaving behind a high-Magnesium high-Potassium soil such as is commonly found in forests in "humid" climates. (Humid in this usage means a climate where the amount of precipitation exceeds the amount of evaporation + transpiration from plant leaves, the evapo-transpiration ratio; it is not referring to the relative humidity of the air.) There are entire communities of plants well adapted to these low Ca, high Mg and K, generally acid soils, e.g. rhododendrons, azaleas, coffee, cacao, holly, blueberries, and many members of the ericaceae family.

Over the eons, as the leaching and acidification process continues, more and more base cations will be lost. The soil becomes more acidic as the negative exchange sites are filled with H+ ions, and eventually Aluminum+++ ions, as $Silicon^{4+}$ is dissolved from the aluminosilicate clay matrix, leaving Al+++ in solution. Finally the soil loses almost all of its exchange capacity, almost all of its ability to hold onto nutrient ions. It becomes a degraded, nutrient-poor highly acidic soil with high levels of soluble Aluminum and other metals that are toxic to plant roots. This type of soil is commonly found in tropical rain forests.

Natural precipitation carries acidic molecules that seek a balance with alkaline elements. If precipitation is higher than evaporation, the acidic precipitation will eventually erode and dissolve all alkaline rocks and minerals. Below pH 7, in most soils, there will no longer be any alkaline compounds that can be readily dissolved, and most of the base cations will either be held on exchange sites, in use by soil biology, or in solution in the soil. In a living soil below pH 7, there will only be very small amounts of Ca, Mg, Fe, etc in solution; the majority of cation elements will be held on exchange sites on clay or humus, or complexed with organic life and decaying organic matter. Soil organisms and plant roots, living and dead, also have exchange capacity.

Many of our best agricultural soils lie in the range of pH 6.0 to pH 7.5 because at those pH levels there is generally an abundance of the nutrient cations Ca, Mg, and K, and the soil is neither so alkaline as to make other elements like P, Fe, Mn, Cu, and Zn poorly available, nor so acid that there are few base cation nutrients available.

Measuring Soil pH: The most accurate way to measure soil pH is with a quality pH meter. pH testing paper may also be purchased from a pharmacy, an aquarium supply store, or a swimming pool supplier. For Ideal Soil purposes the measurement should be taken on a 1:1 ratio of dry soil to water, by weight. Weigh out e.g. 30 grams of dry soil and 30 grams of distilled water. Stir or shake them together, and then let them rest for 1 hour, stirring or shaking occasionally. The pH meter probe (or paper pH testing strip) is then inserted into the mixture and the reading taken.

Back to Soil Tests and Estimating CEC

When we add a **strongly acidic extractant** to a **soil with a pH below 7**, shake it in a test tube for 20 minutes, and then filter off the resulting liquid, most of the base cations Ca, Mg, K, and Na found in the extractant will have come from negative exchange sites. We can measure their amounts, and then, knowing their positive charges, add those charges together and get a rough estimate of CEC. Knowing the soil's measured pH, we can estimate what percentage of the negative exchange sites are occupied by H+ ions and other bases, and come to a close approximation of actual, functional exchange capacity.

However, when a **strongly acidic extractant** is added to a **soil at pH7 or above**, the acid will attack and dissolve base cations from alkaline minerals in the soil, not just from exchange sites.

> **The Fizz Test:** A simple way to determine if a soil has an excess of base cations that could be extracted by the soil testing solution and cause error in estimating CEC is to pour a small amount of ordinary household vinegar on a sample of dry soil. If the soil fizzes and bubbles, there are excess cations and the soil will need another test in addition to the Mehlich 3 test in order to accurately measure exchangeable bases. Most soils below pH 7 will not fizz or bubble, but some soils, especially calcareous sands which are often used on golf courses, can have an overall pH below 7 and still have undissolved limestone particles.

The commonly used multi-element extractants used for soil testing are either acidic or neutral pH. The Mehlich 3 test is pH 2.5, the Mehlich 1 is pH 1.25. The Morgan solution used in the Lamotte test is pH 4.8; the Neutral Ammonium Acetate test is pH 7.0. Obviously, if any of these tests are used on a soil with a higher pH than the test extractant, they can dissolve alkaline elements from soil particles, not only from exchange sites.

Calculating CEC

As we learned earlier in this book, the classic formula for calculating Cation Exchange Capacity is

$$\frac{\text{ppm Ca}}{200} + \frac{\text{ppm Mg}}{120} + \frac{\text{ppm K}}{390} + \frac{\text{ppm Na}}{230} = \text{CEC in meq}$$

The parts per million of Calcium extracted, divided by 200, is our estimate of the amount of Ca being held in exchangeable ionic form. As we wish to know only the amount of base cations held on negative exchange sites, any excess cations extracted will give a false high estimate of CEC.

Our true goal is to **measure the total number of negatively - charged sites that are potentially available** to hold and exchange base cations. There are soil tests that will do just that, but they are complex, expensive to perform, and may contain toxic metals, e.g. the Barium chloride extraction, or reagents that are corrosive to lab equipment, e.g. the Sodium Acetate/Ethanol CEC test.

> William Albrecht determined the exchange capacity of the colloidal clay he used in his BCSR experiments by running a DC current through the clay suspended in water and filling all of the negative exchange sites with protons, i.e. H+, and then displaced the H+ by adding various base cations to the suspension. Measuring the **amount** of base cations **adsorbed when the H+ ions had all been displaced** gave him the total CEC, the total permanent negative electrical charge of the clay fraction he was experimenting with.

For the first century of soil chemical analysis, each element was measured individually through a separate series of steps with different reagents. Often a completely different procedure was used to extract various elements; one extractant and test for Fe, another for Cu, another for B, a time consuming and complex process.

For a routine soil analysis we do not want to involve many separate steps and different tests. Ideally we want a test that will extract all of the elements we wish to measure, and will at the same time allow us to closely estimate CEC.

The two present-day "universal" soil tests that do extract most or all of the exchangeable cations are the Ammonium Acetate and the Mehlich 3 tests. The Mehlich 3 and the neutral ammonium acetate AA 7.0 will both extract about the same amount of Ca, Mg, K, and Na. Both are strong extractants that will not only strip exchangeable cations, but can etch and dissolve carbonates, or any other rock that is more alkaline than they are.

Given the ability of both the M3 and AA extractants to extract more base cations than are actually exchangeable, **which soil test can be used** to extract exchangeable cations and only exchangeable cations from a high-pH soil? The answer turns out to be **the ammonium acetate test,** but only after it has been modified to have a pH higher than the soil sample being tested.

Ammonium acetate is made by mixing aqueous ammonia NH_3 with acetic acid CH_3COOH, the acid found in common vinegar. The pH of the resulting ammonium acetate solution ($NH_4C_2H_3O_2$) will depend on the ratio of acetic acid to ammonia in the mixture. If more ammonia is added the mix becomes more alkaline, more acetic acid makes it more acidic. For soil extractant use the mixture has usually been made at pH7 or pH7.2. Adding a pH7 solution to a soil of pH >7 will result in alkaline mineral compounds being dissolved along with exchangeable bases. By adding more ammonia to the solution, the pH can be raised to 8.0 to 8.5, above the pH of most agricultural soils. For soils in the 7.0 to 8.0 pH range the ammonium acetate extractant is commonly raised to pH8.2. This is known as **the AA8.2 soil test**.

The high concentration of ammonium NH_4^+ in the AA solution readily displaces (exchanges for) the base cations Ca, Mg, K and Na from negative exchange sites. This is due to **three factors governing ion exchange: The relative concentration of the ion, the electrostatic charge of the ion, and the radius (size) of the ion.** NH_4^+ has a molecular weight of only 18 (N=14 H=1), less than Na at at.wt. 23, and a small radius which allows it to fit into small spaces such as between clay layers in expandable clays. In solution, NH_4^+ has approximately the same diameter and charge as K^+. NH_4^+ can fit between clay layers, and when it is in a higher concentration in the soil-water solution than the other cations, will readily replace Ca, Mg, K, Na, and H^+ on exchange sites.

In 1848, Harry Way poured an ammonium solution, ammonium sulfate, through a column of soil. When he analyzed the solution that had leached through the bottom of the soil column, he found it contained less ammonium but high amounts of Ca, Mg, and K. This was the first scientific description of ion exchange.

The AA8.2 Test VS the Mehlich 3 Test

If the AA solution has a higher pH than the soil sample, the acetic acid will be neutralized and will not react chemically with alkaline minerals. The base cations in the resulting soil extract will largely be those that have been exchanged for ammonium NH_4^+, and their amounts can be used to accurately estimate CEC.

Real World Examples

Following are two soil tests reports comparing the results of a Mehlich 3 and an AA8.2 test on the same soil sample. This first example is from the garden of a giant pumpkin grower in Walnut Grove, California.

Location: Walnut Grove, Sacramento County, California

Element	Mehlich 3		AA 8.2 pH		Ideal Soil Ratios	
Cation Exchange Capacity		19.05		10.26		10.26
pH of Soil Sample		7.80		7.80		
Organic Matter %		13.13%				
Sulfur S- ppm		37				K x 0.50 = 100
Boron B ppm		1.36				1.71
Base Cations	%CEC	ppm	%CEC	ppm	%CEC	ppm
Calcium Ca++	59.06%	2256	64.38%	1321	83.50%	1713
Magnesium Mg++	26.12%	597	21.28%	262	10.00%	123
Potassium K+	13.47%	1001	13.40%	536	5.00%	200
Sodium Na+	1.21%	53	0.93%	22	1.50%	35

The first thing to notice is that calculated CEC has dropped 46%, from 19.06meq to 10.26meq, because the AA8.2 test has extracted many fewer base cations than the M3 test, approximately 1/2 as many.This soil also has an excess of K and Mg and a deficit of Ca.

Note that the Ideal Soil ratios in the third column are changed significantly from the "ideals" for a soil with a pH <7. Rather than a ratio for Ca:Mg:K:Na of 68:12:4:1.5 or 70:15:4:1.5, with 10 to 15% exchangeable H+, 100% of the CEC is accounted for by the base cations. Our working **"ideal" BCSR ratio for a soil ≥7 pH is**

83.5% Ca 10% Mg 5% K 1.5% Na

To calculate the desired amounts of P, B, Fe, Mn, Cu, and Zn, we use the amount of K at 5% of CEC. Ideal K for a soil with a CEC of 10.26 is 200ppm, so Ideal P will be 200ppm, Fe 100ppm, Zn 20ppm etc. The amount of Sulfur desired can be higher than K x 0.50 in order to help displace any out of balance cations (the excess K and Mg in this case), from the exchange sites.

If the CEC number calculated from the Mehlich 3 test had been used, the "ideal" amounts of secondary minerals would have been too large. Ideal K at 4% of 19.06meq would be 297ppm, at 5% of 19.06meq, 372ppm.

At first glance this soil appears to have enough Ca potentially available. The M3 test found 2256ppm Ca, more than enough to bring the Ca up to the desired 1713ppm (83.5%), but apparently that Ca is not soluble or available in this soil, so we would want to raise the level of soluble Ca in the soil with the aim of replacing some of the excess exchangeable K and Mg with Ca.

The Soil Rx written for the Walnut Grove soil called for 1525# of gypsum $CaSO_4$. At 22%Ca and 16% S, 1525# = 336#Ca and 224#S, or 168ppm Ca and 112ppm S added to the plow layer.

According to the Ideal Soil Chart, B should = Ideal Ca x .001. Boron in the Walnut Grove soil was measured at 1.36ppm by the M3 test. The Rx called for 10# of borax 9%B, which adds 0.45ppm B and raises the total B to 1.81ppm, a little more than 1/1000th of our desired Ca saturation of 1713ppm.

Example 2: Mehlich 3 vs AA8.2, Center Point Texas

The next example is a limestone-derived soil that is common in central Texas. The grower wanted to put in a new USDA Organic orchard and vegetable garden. The Mehlich 3 test extracted 13815ppm Ca. 13815 / 200 = 69.08meq of Ca. The AA8.2 test found only 1932ppm of exchangeable Ca. 1932 /200 = 9.66meq of Ca.

Total CEC dropped 84% from 71.35 to 11.15.

Location: Center Point, Kerr County, Texas

Element	Mehlich 3		AA 8.2 pH		Ideal Soil	
Cation Exchange Capacity		71.35		11.15		
pH of Soil Sample		7.60		7.60		
Organic Matter %		4.06%				
Sulfur S- ppm		20			K x 0.50 =	109
Boron B ppm		1.02				2.0
Base Cations	%CEC	ppm	%CEC	ppm	%CEC	ppm
Calcium Ca++	96.81%	13815	86.64%	1932	83.50%	1862
Magnesium Mg++	1.34%	115	3.89%	52	10.00%	134
Potassium K+	1.73%	482	8.78%	382	5.00%	217
Sodium Na+	0.12%	19	0.74%	19	1.50%	38

Unlike the Walnut Grove soil, the Center Point soil has a very high level of Ca and a good reserve of K but is seriously lacking in Mg. The aim should be to increase the level of Mg and displace some of the excess Ca from the exchange sites. If the M3 test estimated CEC of 71.35 had been used to make the calculations, Ideal Mg at 10% of CEC would have been 856 ppm rather than the 134ppm Mg "ideal" listed above.

The Rx written for the Center Point garden and orchard called for 2180# of Magnesium sulfate $MgSO_47H_2O$ (Epsom salt). At 10%Mg and 14%S, that would add 218# of Mg and 305# of S, or 109ppm Mg and 153ppm S, a little more than the amount needed to raise Mg to 134ppm. The next pair of soil tests after the present growing season will tell if that was enough Mg to reach the desired target.

Magnesium sulfate is a more costly and less concentrated source of Mg and S than Magnesium Oxide and agricultural Sulfur, but was used in this case because Magnesium oxide is not allowed under USDA NOP rules.

10# of Solubor 20% B was also recommended, adding 2# or 1ppm B to the topsoil. This is a high Calcium soil and we wish to have optimum levels of B available. There would be little danger of adding too much Boron to this soil with the extremely high level of Calcium it contains.

Calcareous Soils with pH < 7.0

As mentioned above, some sands have developed partially from eroded limestone. A soil largely composed of these calcareous sands, in a humid climate or when heavily irrigated as on lawns and golf courses, may have a measured pH below 7 but contain significant amounts of undissolved Calcium or Magnesium carbonates. These can be extracted by a Mehlich 3 or other acidic soil test and give a false high estimate of CEC. If one suspects they are dealing with a soil like this, the "fizz test" described above, where ordinary vinegar is added to the soil, is recommended. If the soil bubbles when vinegar is added, you will want to get an AA8.2 test in addition to the essential M3 test.

Another indicator that one should get an AA8.2 test for a soil with pH <7 is if the Mehlich 3 test shows a surprisingly high level of Calcium and estimated CEC.

Calcareous sands will generally have a low CEC, often as low as 2 or 3meq, when tested with the AA8.2 test. If the AA8.2 test shows a CEC >7meq and a pH <7, they may be treated the same as any other soil with a pH <7, according to the ratios shown on the Ideal Soil Chart. If they have a CEC < 7meq, they should be treated as shown in the following chapter on Low CEC Soils.

Mineral Availability in High pH Soils

The availability of essential minerals such as Fe, Mn, Cu, Zn, and P can be poor in high pH soils because the metallic elements readily combine with OH- ions, forming insoluble hydroxides, while P combines with available cations forming insoluble phosphates. Often this problem is addressed by applying these elements in soluble form as sulfates, chlorides, or nitrates via fertigation, foliar spraying, banding of soluble fertilizers in the seeding row, or by using small amounts of the elements chelated with EDTA.

In the 4th century BC Aristotle wrote that "the soil is the stomach of the plants, digesting and making food available". As in our own bodies, it is the beneficial microorganisms that do much of the work of digestion and chelation of essential minerals.

One significant benefit of having the minerals in balance in the soil is that it encourages the growth of beneficial soil organisms that can access poorly soluble nutrients and chelate them biologically. As the soil organisms go through their life cycle, the minerals they have chelated become available to plants. Mycorrhizal fungi have been shown able to achieve a pH<2, which can solubilize practically any mineral complex in the soil and make it available.

It has been our experience that if the anions N, P, Cl, and especially S are maintained at Ideal Soil ratios, along with a optimum level of organic matter in a biologically active soil, the metallic elements and P will remain available to the crops without resorting to spoon-feeding.

Chapter 10

Working with Low-CEC Soils (Below 7 meq)

Minimum Amounts Needed for Ideal Soil Method in Any Soil or Growing Media
Based on results from a Mehlich 3 soil test

Calcium Ca	**750 ppm (1000 ppm is better)**
Magnesium Mg	**100 ppm**
Potassium K	**100 ppm**
Sodium Na	**25 ppm**
Phosphorus P	**100 ppm**
Sulfur S	**50 ppm**
Boron B	**1 ppm**
Iron Fe	**50 ppm**
Manganese Mn	**25 ppm**
Copper Cu	**5 ppm**
Zinc Zn	**10 ppm**

A soil with a CEC below 7meq does not have sufficient negatively charged exchange sites to adsorb and hold onto the minimum amount of nutrient cations necessary to achieve Ideal Soil results. Nonetheless, the minimums are needed, even if the soil can't hold them against leaching.

A sandy or leached-out, low-organic matter soil may have a CEC of 3meq or less. With a CEC of 3meq, at 100% saturation, the soil could adsorb and hold a maximum of

Calcium: 200ppm x 3meq = 600ppm Ca
or
Magnesium 120ppm x 3meq = 360ppm Mg
or
Potassium 390ppm x 3meq = 1170ppm K
or
Sodium 230ppm x 3meq = 690ppm Na
(See Chapter 2, Cation Exchange)

If we try to balance the primary cations in a CEC 3meq soil to a **standard 68%Ca : 12%Mg : 4%K : 1.5%Na cation saturation ratio**, we end up with only 408 ppm Ca, 43ppm Mg, 47ppm K, and 10 ppm Na, far below the Ideal minimums.

The short-term solution is to apply the amount of minerals needed to achieve the Ideal minimums, realizing that some part of what is added will not be adsorbed and held to an exchange site, and will be subject to leaching. Irrigation in a low CEC soil should be kept to a minimum, adding only what the soil can hold. If there is excessive rain, it's safe to assume much of what was added (other than Phosphorus) will be leached out and will need to be added again. This may call for frequent soil tests and amending if an optimum yield with maximum nutrient content is desired.

Note that even though a large amount of base cations may need to be added to achieve the minimum amounts of Ca, Mg, and K, as long as the concentration of anions (P, S, NO_3, Cl) is also at optimum, the soil's **H+ : OH-** balance will settle out at around pH 6.5

Increasing the Exchange Capacity

The longer-term answer is to increase the exchange capacity of the soil. This can be done by adding or increasing.

Soil Organic Matter (SOM), as stable humus
Humate ores
Charcoal or Biochar (e.g. terra preta soils)
Low-fired pottery sherds (also found in terra preta soils)
Expanded Vermiculite
High-CEC clay such as Calcium bentonite/montmorillonite clay
Zeolites

Adding or Increasing Organic Matter

Stable humus has a CEC of up to 200meq and significant anion exchange capacity as well. Organic matter in the process of breaking down doesn't have much exchange capacity; it only achieves that when it is broken down to stable humus that can no longer readily serve as food for plants and soil organisms. Organic matter decomposing into stable humus will only happen efficiently in a biologically active, mineral balanced soil with optimal levels of N, C, S, and Ca. Even if the **Carbon:Nitrogen ratio** of the decomposing organic matter is **an ideal 25 or 30:1**, without sufficient S to form Sulfur-containing amino acids like cysteine and methionine, Nitrogen released during decomposition can be lost to the atmosphere as ammonia NH_3 gas, or form water-soluble nitrate NO_3 and be leached away. Likewise without sufficient Ca^{2+} to bind with the CO_3^{2-} carbonate ions produced during decomposition (forming stable Calcium carbonate $CaCO_3$), the Carbon released may be lost to the atmosphere as CO_2.

Only with the proper mineral balance in the soil (or compost pile) will the maximum amount of stable humus with optimum exchange capacity be formed and conserved.

The strategy of building humus to increase exchange capacity **only works well** in cooler temperate climates **where the annual precipitation is equal to or greater than the average annual amount of precipitation that evaporates from the soil or transpires from the plants** (called **the evapo-transpiration ratio or rate**). In warmer tropical and sub-tropical climates the biological activity is high and the decomposition process so rapid that it becomes difficult to increase the humus content of the soil on a long-term basis.

In any event, what the climate dictates naturally will be the easiest level of organic matter and humus to maintain; this is generally the same level as will be found in an undisturbed forest or prairie of a similar soil type in a given climate. As a rule, the further away from the Equator one goes, or the higher in altitude at a given latitude, the higher the level of humus and organic matter in the soil will be. It can be difficult to maintain even 2% Soil Organic Matter (SOM or just OM) in tropical lowlands, while 10% or more SOM is common in some humid boreal climates. Bog and peat soils are also found in tropical and subtropical climates but their high organic matter content soon burns up when put into cultivation.

Humate Ores and Humic/Fulvic Acids

Humate ores form from massive layers of vegetation that have either been covered over by subsequent sedimentary layers such as sand or volcanic ash, which diminish leaching, or they form in impermeable clays below beds of soft coal, where, as the humic substances leach out, they may be stopped by, held, and concentrated in an underlying clay or shale layer (if one exists).

When a large area of this ancient vegetation has been sealed in by a sedimentary layer of stone above it, it will not leach but continue to percolate, rising to the stone lid above it, condensing, trickling back down, over and over for millions of years. What effect this process may have on the properties of the humic substances is unknown, but has been well described in the old Alchemical literature:

[...]Now in our Art you should closely imitate these natural processes. There should be the Central Heat, the change of the water into air, the driving upward of the air, its diffusion through the pores of the earth, its reappearance as condensed but volatilized water.
The New Chemical Light, by Michael Sendivogius, 17th Century

[...] The earth conceives in her womb the splendour of the Sun,and by it the seeds of the metals are well and equally warmed, just like the grain in the fields. Through this warmth there is produced in the earth a vapour or spirit, which rises upward and carries with it the most subtle elements. It might well be called a fifth element for it is a quintessence, and contains the most volatile parts of all the elements. This vapour strives to float upward through the summit of the mountains, but, being covered with great rocks, they prevent it from doing so: for when it strikes against them, it is compelled to descend again. It is drawn up by the Sun, it is forced down again by the rocks, and as it falls the vapour is transmuted into a liquid, i.e., sulphur and mercury. Of each of these a part is left behind—but that which is volatile rises and descends again, more and more of it remaining behind, and becoming fixed after each descent.
The Glory of the World, Or, Table of Paradise, by Anonymous, 1526 AD

Top quality humate sources will have a humic acid content above 70% by weight and an exchange capacity of 200-300 meq. Application rate varies depending on soil type, but in low CEC soils around 400 lbs/ac or 400 kg/ha annually would be our recommended maximum application.

Charcoal or Biochar (Terra Preta Soils)

Charcoal made from plant matter, more or less finely ground or powdered, and intended as a soil amendment, is also called biochar. Charcoal or biochar has one of the highest exchange capacities of any known material.

> *One of Biochar's secret assets lies in its large surface area of approximately 500 m2 per gram! This is as a result of the micropores formed during pyrolysis. In general the higher the pyrolitic temperature the larger the surface area of the finished material until it reaches a temperature at which deformation occurs. It is this large surface area that provides a vast protective habitat for beneficial bacteria and fungi in soils where it is applied; a sort of coral reef for Fungi and Bacteria.*
>
> *Another important attribute is the Cation Exchange Capacity (CEC) of Biochar. CEC is a measure of the surface charge in a soil or a Biochar and in basic terms it is the ability of a soil/Biochar to hold onto nutrients. The benefits for soil work both ways as it will absorb* [sic] *nutrients and prevent leaching yet release the nutrients when required.*
> *http://www.biocharireland.com/science.html*

Researchers have measured the CEC of "fresh" biochar made from pine sawdust pellets and pine timber ranging from 22meq to 138meq. (Characterization and Comparison of Biochar, Herbert et al, CalPoly2012). It is also known that as biochar ages its exchange capacity can increase, up to an order of magnitude (10x).

In 2006 researchers compared several ancient char-amended soils (terra preta androsols) in the central Amazon with adjacent soils to which char had not been added. The most impressive result was an androsol with an Effective CEC of 213meq compared to adjacent soil with an ECEC of 23meq. This same androsol, estimated to be 600 to 1000 years old, tested as containing 9064ppm Phosphorus and 17 545ppm Calcium, vs the adjacent soil with only 273ppm P and 115ppm Ca. (Black Carbon Increases CEC in Soils, B Liang et al, Soil Sci. Soc. Am. J. 70:1719–1730, 2006)

The advantages of using biochar to increase exchange capacity include:

- Unlike organic matter and humus, which can quickly decompose, especially in warm climates, biochar is stable for centuries or millennia.
- Unlike humate ores, biochar can be used in large amounts.
- Biochar can be made from free or inexpensive local waste materials and produced nearby or on site.

The potential disadvantages of biochar are generally short-term:

- Freshly made biochar is hydrophobic and difficult to wet.
- Biochar may take a number of years to reach full exchange capacity.
- Freshly applied biochar can lower the soil concentration of essential nutrients via simple dilution.
- Fresh biochar can strongly adsorb large amounts of nutrient anions and cations from the soil/water solution, making them less available to plants and soil organisms, at least temporarily.

Significant research work is ongoing with biochar. One promising avenue is using it as a carrier for soluble fertilizers ranging from fish emulsion to ammonium sulfate. **Biochar should not be considered a source of plant nutrients.** Adding biochar alone will in general only add a modest amount of Potassium. Its main value is to retain nutrient ions in soils that lack exchange capacity; its important secondary values are its ability to hold water, loosen soil texture, and provide habitat for soil organisms.

Low-Fired Pottery Sherds

Terra-cotta type low-fired pottery sherds are also reported as abundant in terra preta soils. These highly-porous lightly-fused clays have been utilized for their ability to capture and hold onto elements long before the concept of exchange capacity was known: Ancient sailors reportedly used terra-cotta jugs to desalinate sea water. If a sealed, empty jug was held under water, the pure H_2O would filter to the inside while the salt and other elements were adsorbed and held by the negative charges in the terra cotta. The actual exchange capacity of terra cotta pottery sherds is unknown but perhaps worthy of further investigation.

High-fired clays such as stoneware and porcelain are non-porous and would not be expected to add any CEC to a soil.

Expanded Vermiculite

Vermiculite is a hydrous mica, a type of collapsed 2:1 alumino-silicate layer clay where the space between the clay layers is filled with hydrated Magnesium ions. When vermiculite is heated, the water molecules surrounding the Mg^{++} ions turn to steam and the spaces between the layers expand greatly, making a large surface area available for cation exchange. Expanded vermiculite has a CEC of ~100meq per 100 grams.

Expanded vermiculite is very light-weight and has excellent water-holding properties as well as high CEC. It is primarily used in potting soils and greenhouse beds. Its main disadvantage is that when fresh its exchange sites are saturated with Magnesium and Iron, which can add undesired high levels of Mg and Fe to the soil. When expanded vermiculite is used, the growing media should be analyzed for Mg and Fe content, and the amount of vermiculite adjusted so as not to overload the Mg saturation %. The Calcium level must also be carefully adjusted to maintain the desired Ca:Mg ratios.

High CEC Clays: Calcium Bentonite

Calcium bentonite is an amendment that only needs to be brought to an optimal level in the soil one time; its exchange capacity and physical properties will then remain in the soil permanently unless eroded away.

Bentonites are hydrated 2:1alumino-silicate layer clays. Structurally they consist of alumina sheets sandwiched between SiO_2 tetrahedral layers. Unlike the vermiculite micas, bentonite clays readily hydrate and expand in their natural state. Of the two main varieties, Calcium and Sodium bentonite, Calcium bentonite is the one used for agriculture. Sodium bentonite contains excessive amounts of Na, attracts and retains too much water, and swells and shrinks excessively. Calcium bentonites typically have a CEC of 70 to 100meq.

Calcium Bentonite is is an excellent amendment for loose, low-CEC “beach sand” type soils. The Ca bentonite will disperse and coat the sand grains, helping them stick together to form aggregates. Calcium bentonite also interacts chemically and biologically with humus and decomposing organic matter, preserving it the soil and providing habitat for soil microbes whose excretions further bind the sand grains and give the soil a cohesive structure while adding much-needed CEC.

Calcium bentonite is also a significant source of exchangeable Ca, Mg, K, and Na. It **may be used in any sand or silt soil** or in growing media where readily available Ca, increased CEC, and better moisture and fertilizer retention are desired. **For heavy clay soils with low CEC,** where adding more dispersed clay is generally not desired; other CEC builders like **biochar, humates, granular zeolite, and increased SOM are more appropriate.** See below.

Calcium bentonite works best when combined with mature compost, or mixed in at the start of the composting process, or when used in soils with optimum humus and organic matter levels.

Application rates for Calcium Bentonite:

Loose sandy soils: 300 to 600 kg/are (600 to 1200 lbs/1000 ft^2) mixed into the top 15 to 30 cm (6 to 12 inches)

Established gardens and lawns: 100 to 200 kg/are (200 to 400 lbs/1000 ft^2). If bentonite cannot be tilled in, the application should be split into three or four doses applied separately and raked and watered in. The clay will leach into the soil and find its proper place. It may also be mixed with water at a 1:10 ratio by volume and applied with a watering can.

Zeolites

Clinoptilolite (a naturally occurring hard clay compound with high CEC) is the zeolite generally used in agriculture. It is mined as a hard clay compound that is then crushed to a granular form. Unlike Ca bentonite, zeolites remain in granular form and do not disperse in the soil. They may be used in any soil type, but are especially useful for loosening and adding CEC to heavy low-CEC clay soils such as the kaolin clays common in the tropics and subtropics. They may also be put to good use in potting mixes.

Natural zeolites form where fresh volcanic lava reacts with sea water, and where volcanic ash layers react with alkaline groundwater. Zeolites also may form in shallow marine basins. What makes a zeolite a zeolite is its very regular, rigid, open latticework crystal atomic structure with a network of interconnected pores. Zeolites may have a surface area of up to 450 meters^2/gram and as much as 45% open pore space for holding water and exchangeable cations.

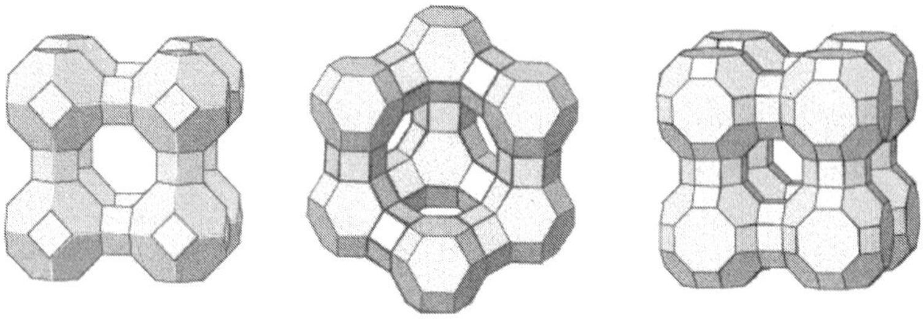

2. Crystal structure of zeolites. Note the pyramidal structure of Clinoptilolite, center.

Exchangeable Minerals in Zeolites

Zeolites that form from volcanic lava in contact with sea water, and those formed in other saline environments, may have a high content of Sodium, though often this has been displaced over time, largely by Potassium. One should know the typical mineral analysis of any zeolite they are using in agriculture.

Restoring Aged Clays

In some cases the soil may have a higher potential exchange capacity, but the clays it contains consist of collapsed clay layers that have been filled with K, as Ca and Mg leached out over millennia, or have been abused by excessive applications of Potassium chloride KCl fertilizers that "plug up" the exchange sites on the edges of layered clays, or fill the spaces between the layers with K+. Potassium just happens to be the perfect size to collapse the space between clay layers so that Ca and Mg can no longer fit in between to attach to the (**–**) negative exchange sites. To restore a collapsed layer clay or one where the exchange sites on the edges are plugged with K, the logical approach would be to greatly increase the Calcium saturation in the soil with the aim of displacing K from the layered-clay edges, and perhaps even expanding the layers that are presently collapsed by displacing some of the interlayer K with Ca.

Gypsum, Calcium sulfate, is the best tool for opening collapsed layer clays or displacing excess cations because it readily dissolves and dis-associates into Ca^{2+} and SO_4^{2-} ions in the soil/water solution. A high enough concentration of Ca ions will tend to displace any other cation on the exchange sites; the displaced ions will then either be taken up by plants and soil organisms or quickly associate with free SO_4^{2-} forming K_2SO_4, $MgSO_4$, Na_2SO_4 etc. which can readily be leached to a lower soil horizon.

A Review of Methods and Materials for Increasing Exchange Capacity

Increasing the soil's humus content and thereby its exchange capacity by growing and decomposing organic matter right in the soil is generally the least costly method pf increasing exchange capacity. Given the proper amount and balance of mineral nutrients, humus will be formed and conserved, but the level of humus that can be maintained easily will be dictated by the climate. In a cool, rainy climate, maintaining `above 5% SOM is easy; in humid tropical or warm desert climate it is an endless struggle, even with large amounts of organic matter brought in from off site. All growers should do their best to maintain optimum SOM and humus levels in the soil, but that may not be enough to raise the CEC to the desired level..

Humate ores can add some exchange capacity, but are limited in the amount that can be safely applied, generally 200 to 300 ppm annually. Presently it is unknown if there will be a significant cumulative effect on exchange capacity from long-term application of humates.

Expanded Vermiculite is an excellent source of CEC, but has **a high amount** of **Magnesium** and often **Iron** already adsorbed to the exchange sites. The amount of Mg and Fe being added must be taken into account when vermiculite is used. Expanded vermiculite is relatively expensive and normally used only for potting mixes and container growing media.

Perlite is a naturally occurring volcanic glass that contains water. When heated above 850° C the glass softens and the water expands the perlite to 7 to 16 times its former volume, making a lightweight, porous, inert glass bead. **Perlite has almost no exchange capacity and is not useful for raising CEC.**

High CEC clays may naturally occur not far from sandy or gravelly soils with low CEC, and it may be economical to dig, haul, spread and mix them into a low-CEC soil. One should first make certain that the clay being considered as an amendment does have a high enough CEC to be worth the effort. Many sticky, dense clays such as kaolin have a very low CEC. The only way to know is to have them tested. For a soil with pH less than 7, a Mehlich 3 test would be appropriate; for clays with a pH above 7, the Ammonium Acetate pH8.2 test is better. **See Chapter 9 on Calcareous and High-pH Soils for more info on the AA8.2 soil test.**

Calcium Bentonite is an excellent CEC booster for soils that don't already have a high clay content, but large amounts must be used to make much difference. An application rate of 300 to 600 kg/are (600 to 1200 lbs/1000 ft^2) works out to 30 to 60 metric tons per hectare or 13 to 26 avoirdupois tons per acre. Adding those amounts will only be economical if there is access to an inexpensive source of Ca bentonite and transportation costs are not too high.

Pot sherds, broken low fired pottery, may be worth investigating if one has access to a sufficient quantity nearby that is free or low cost.

Biochar/charcoal is likely to end up being the most economical way of adding exchange capacity to a soil, especially after the "bugs" are worked out of the system and we learn how to make top-quality high-CEC and high-AEC biochar efficiently in large quantities.

Which or what combination of the exchange capacity amendments above should be used will depend on cost, availability, existing soil type, and the personal preferences of the grower.

Epilogue: The Three-Legged Stool

The most stable of constructions is the tripod, a word that comes to us from ancient Greece and means, precisely, "three-footed".

Organic life, mineral elements, and energy are the three components of our 3-dimensional world; together they are the "legs" of the three-legged stool that supports our agriculture, as well as everything else we see, touch, hear, smell, and eat. Let's look at them a little more deeply.

Organic life: Whatever forms the life force animates on this planet are Carbon-based. Living things use Nitrogen to form proteins and amino acids; their essential solvent and lubricant is water, H_2O, made up of Hydrogen and Oxygen. These four elements, C,N,H, and O, are the air elements; they come from the atmosphere and return there. Plants use them to make their living forms, animals eat the plants, both plants and animals grow, reproduce, and die, and their constituent compounds of C,N,H, and O cycle endlessly through the biosphere. Their dead and decomposing remains make up the organic fraction of the soil, from recently fallen leaves and branches to humus and the humic and fulvic acids that are the final product of their breakdown. These Carbon-based life forms comprise that part of the soil that has gotten so much attention from the proponents of alternative and sustainable agriculture over the past sixty years. Great strides in understanding have been made, but the organic fraction and soil life are still only one leg of the tripod; this leg alone will not support the New Agriculture.

Energy is the motive force; it provides power to the growth/decay process. We are living in a sea of energy as surely as fish swimming in the ocean are living in water. Light, sound, and electromagnetic energies surround us always. Our heart beats from an electrical signal; our nerves fire electrically; chemical reactions result from the exchange of + and - electrical charges. There are also more subtle energies that we don't yet have the ability to measure and quantify. What is telling the DNA to do what it does? We call it the life force, but science has no clue what it is. Whatever it is, it is directing the form that living things assume, from opossums to sunflowers to re-growing an injured fingertip. When the life force goes away, the form breaks down and becomes food for other living things, on down the chain, until its components eventually end up as minerals back in the sea and elements in the air. The energy of the life force directs the form, growth, and decay in the complete cycle from mineral back to mineral.

If one loves and praises their garden or farm and all of the life therein, Nature responds and the plants thrive. The old saying goes that the best fertilizer is the farmer's footsteps, walking through his fields. What subtle energy is this? There are also the moving, changing magnetic fields of the Earth, interacting with magnetic and paramagnetic elements in the soil. Why do cabbages grow to weigh fifty pounds in Alaska? Is it because of the long summer days, or does it have something to do with the magnetic lines of force being closer together?

The analogy of the soil to an electric storage battery is still a good one: What is the difference between a fully charged battery and one that is stone-cold dead? The chemical composition of both is the same, but the charged battery has a potential difference between the terminals; when they are connected chemical reactions begin to occur and energy flows.

Carey Reams and Rudolph Steiner both had a strong appreciation of the importance of energy flow in Nature; at the times they were teaching these were new concepts in agriculture and there were no accepted terms with which to describe these energies; there are still no accepted terms today. Phil Callahan's groundbreaking work with paramagnetism is still little known at present, though the hard empirical evidence of its effectiveness and results is undeniable. A solid understanding of and coherent explanation of energy in agriculture remains the least understood leg of our tripod.

Mineral Elements: The building blocks that the life force directs and energy powers are the physical elements from Hydrogen to Uranium. These elements are what everything is made out of. The elements themselves are stable arrangements of protons and neutrons, making up the nucleus, surrounded by negatively charged electrons. The key is that they are stable. Except in the case of radioactive elements of very high atomic weight, they remain what they are; one element does not change into another. Plants and animals require a wide range of elements to do many different tasks. A stalk of wheat requires large amounts of Silicon to keep it stiff and strong; it also needs Copper to give it flexibility and resilience. The germ of every seed needs an atom of Manganese to be fertile, but it also needs Calcium, Boron, and probably Phosphorus to bring the atom of Manganese to where the seed will form. In order for a living thing to make an amino acid, enzyme, or protein a mineral element must be present to act as a catalyst. We know of over 300 enzymes in the human body alone that require Zinc as a catalyst.

The formation of each separate DNA codon requires a mineral catalyst, some of them extremely rare, elements such as Yttrium and Scandium. Each of these must be in the soil if they are to be in our food. Any necessary element that is missing or out of balance will affect the health of the entire system.

These are the legs of our three-legged stool. In order to sit level and support the plants, the soil life, and all life dependent on the soil, all three legs had best be equally long and equally strong. In this book we have shown how to apply the knowledge of mineral balance to create your own "ideal soil" wherever you may be, with the assurance that the mineral leg of the stool will be strong and support its intended load. It will be up to all of us in the future to gather the knowledge and experience from all disciplines and combine them wisely to ensure that the New Agriculture stands on a solid, balanced, and secure foundation.

It is hoped that the reader will take the knowledge gained from this book and apply it to their own situation, combine it with their other knowledge and experience, and share it with others. Together we can create the most magnificent, healthy, sustainable, beautiful planet imaginable, one on which we can all live a healthy and joyful life, and one that we can pass on to future generations with pride.

We can do this.

Index to Appendices

Instructions for taking a soil sample:

You only need to send 1 cup of soil, approximately 8 ounces (225 grams).

The goal is to get a representative sample of your soil 6" deep. This is the aerobic zone where most root growth and nutrient exchange happens. You will want to take several samples of the area and mix them together. As a rule, you would take six to ten samples of any area that you are going to treat the same, while avoiding strange or unusual areas or sampling them separately for a separate test. Clear the sampling area of loose duff and organic matter, mulch etc. before you take your sample. If you have access to a soil probe, which takes a 1" circular sample (a tube of soil) use that. Otherwise just dig a straight-sided hole six to eight inches deep with an ordinary shovel, then take a thin slice straight down along one side of the hole. This slice is your soil sample. Use a clean shovel to do the digging. You don't want to contaminate your soil sample with rust or with dirt from another area. Avoid sampling areas that have had fertilizer applied in bands.

Take as many samples as you think necessary and mix them together thoroughly in a glass, ceramic, plastic or stainless steel container, removing any roots or large chunks of organic matter, then take about one cup of this mixed sample to send to the soil testing lab. If the soil is extremely wet when you take the sample, spread it out in a warm place and let it dry until it can be handled without leaving mud on your hands; it does not need to be totally dried out, just not wet.

Which Soil Test Do You Want?

The Ideal Soil Method was developed and is designed to work best with the Mehlich 3 test. If you are testing a calcareous soil or a soil with a pH above 7.0, you will also want the AA8.2 test for an accurate assessment of exchangeable cations in order to calculate CEC.

Which Soil Testing Lab Should You Send the Sample To?

Most importantly, you will want **a lab that uses the Mehlich 3 test regularly** and that also **offers the AA8.2 test**. These are the elements you need to have data for:

Primary Cations	Primary Anions	Secondary Elements
Calcium Magnesium Potassium Sodium	Phosphorus Sulfur	Boron Iron Manganese Copper Zinc

In addition you will need data for **soil pH and %OM or SOM,** soil organic matter. Be sure to specify that you want the lab to show the **results in parts per million ppm**.

There are many good soil testing labs around the world. Soilminerals.com usually sends their samples to Logan Labs, in Ohio, USA www.loganlabs.com Logan labs sends reports in the same format as used in this book. Soil samples being sent from one country to another will require a soil import permit that can be obtained from the lab.

Contact the soil testing laboratory before you send the soil sample for prices and any special packaging instructions.

The Ideal Soil Chart (Agricola's Best Guess v 2.0 January 2014)

Based on a Soil Test using the Mehlich 3 method

Organic Matter (OM)	2% — 10%	Depending on climate
pH	6.4 – 6.5	Balance the minerals and pH will take care of itself
Primary Cations as % of Cation Exchange Capacity (CEC) **See appendix "Calculating TCEC" p 125**		
Calcium (Ca)++ min 750ppm	60% — 85% (Ideal 68%)	Ca & Mg together should add to 80% of exchange capacity in most agricultural soils pH 7 and lower
Magnesium (Mg)++ min 100ppm	10% — 20% (Ideal 12%)	
Potassium (K)+ min 100ppm	2% — 5% (Ideal 4%)	See Phosphorus (P)
Sodium (Na)+ min 25ppm	1% — 4% (Ideal 1.5%)	Essential for humans and animals
Hydrogen (H)+	5% — 10% (Ideal 10%)	A lone proton. The "free agent"
Primary Anions		
Phosphorus P- min 100ppm	P = **Ideal** K by weight (ppm) **BUT: phosphate** (P_2O_5) should be **~2X potash** (K_2O)	Needs a highly bio-active soil to keep it available.
Sulfur S - - min 50 ppm	1/2 x **Ideal** K up to 300 ppm	Need for Sulfur amino acids Conserves soil N and Carbon
Secondary elements		
Iron(Fe) + min 50ppm Manganese(Mn) + min 25ppm Zinc (Zn) + min 10ppm Copper (Cu) + min 5ppm	Fe: 1/3 to 1/2 x **Ideal** K Mn: 1/3 to 1/2 x Fe **Zn: 1/10 x P** (up to 50ppm) Cu: 1/2 x Zn (up to 25ppm)	Iron and Manganese are twins/opposites and synergists, as are Copper and Zinc.
Boron $B^{3+ \text{ or } -}$ (cation or anion) min 1ppm	1/1000 of Calcium (max 4 ppm)	Essential for Calcium utilization. Calcium transports sugars
Chlorine (Cl)- min 25ppm	1x to 2x Sodium	Essential, but ages clays rapidly when used in large amounts
Silicon $Si^{4 + \text{ or } -}$ (cation or anion)	Ideal unknown. Si is the most abundant mineral in most soils. Active soil biology and balanced mineral chemistry will ensure availability.	
Micro (trace) Elements		
Chromium Cr- Cobalt (Co)+ Iodine (I)- Molybdenum Mo- Selenium (Se)- Tin (Sn)+ Vanadium (V) + Nickel (Ni) + Fluorine (F) –	All of these are essential in small amounts. 0.5 - 2ppm is enough. Some of the micro elements (e.g. Mo, Se) can be toxic to plants and soil organisms in quantities above 1-2ppm. Use Caution when applying micro/trace elements in purified forms	There are probably 30 or so other elements needed to grow fully nutritious food. Sources are amendments such as seaweed, rock dust, ancient seabed or volcanic deposits, rock phosphate, greensand etc

Plants need at least 17 of the 23 elements listed above, as well as Nitrogen, Carbon, Hydrogen, and Oxygen.

Typical Mineral Content of USDA Organic Fertilizer Ingredients (%)

Animal Source	N	P as P_2O_5	K as K_2O	S	Ca	Mg	Fe	Tr
Fish Bone	4	20		0.6	19	0.3		Tr
Fish Meal	10	4.5		0.6	2.3	0.3		Tr
Crab Shell	3	3.25	0.3	0.2	23	0.3		Tr
Blood Meal	13	1						
Feather Meal	12	0.1	0.4	0.4	0.6	0.6		
Bone Meal	3	15			20	0.4		

Mineral Amendments and Kelp

	N	P as P_2O_5	K as K_2O	S	Ca	Mg	Fe	Tr
Ag Lime					32-40	1-5		
Dolomite Lime					22	13		
Gypsum*				16	22			
Oyster shell					36	0.3		
Epsom salt**				14		10		
Potash sulfate**			51	17.5				
TN brown phos		3 (23% total)			40			Tr
Calphos		3 (20% total)			20			Tr
K Mag*			22	22		11		
Greensand		1	7		1.3	2.2	9	Tr
Kelp Meal	1	0.7	3	2	2	0.7		Tr

Tr = Good source of micro (trace) minerals

Purified Source	Sulfur S	Boron B	Iron Fe	Mang. Mn	Copper Cu	Zinc Zn
Ag Sulfur	**90-100**					
Borax**		**9**				
Solubor™**		**20.5**				
Fe sulfate $1H_2O$	**18**		**30**			
Fe sulfate $7H_20$**	**11.5**		**20**			
Mn sulfate $1H_2O$*	**19**			**32**		
Cu sulfate $5H_2O$**	**12.5**				**25**	
Zinc sulfate $1H_2O$*	**17**					**35**
Zinc sulfate $7H_20$**	**11**					**22**

****Highly soluble in H_2O** ***Varies in solubility in H_2O**

Soil Report and Comments

<table>
<tr><th colspan="2">Element</th><th>Results</th><th>Comments</th></tr>
<tr><td colspan="2">Cation Exchange Capacity CEC meq</td><td></td><td></td></tr>
<tr><td colspan="2">pH of Soil Sample</td><td></td><td></td></tr>
<tr><td colspan="2">Organic Matter %</td><td></td><td></td></tr>
<tr><td colspan="2">Primary Anions</td><td></td><td></td></tr>
<tr><td colspan="2">Sulfur S (parts per million ppm)</td><td></td><td></td></tr>
<tr><td colspan="2">Phosphorus P ppm</td><td></td><td></td></tr>
<tr><td colspan="2">Primary Cations</td><td></td><td></td></tr>
<tr><td>Calcium Ca++
ppm</td><td>Desired
Found
Deficit</td><td rowspan="2"></td><td rowspan="2"></td></tr>
<tr><td colspan="2">Ca Base Saturation 60-70 %</td></tr>
<tr><td>Magnesium Mg++
ppm</td><td>Desired
Found
Deficit</td><td rowspan="2"></td><td rowspan="2"></td></tr>
<tr><td colspan="2">Mg Base Saturation 10-20 %</td></tr>
<tr><td>Potassium K+
ppm</td><td>Desired
Found
Deficit</td><td rowspan="2"></td><td rowspan="2"></td></tr>
<tr><td colspan="2">K Base Saturation 2-5 %</td></tr>
<tr><td>Sodium Na+
ppm</td><td>Desired
Found
Deficit</td><td rowspan="2"></td><td rowspan="2"></td></tr>
<tr><td colspan="2">Na Base Saturation 1-5 %</td></tr>
<tr><td colspan="2">Other Bases</td><td></td><td></td></tr>
<tr><td colspan="2">H+ Exch Hydrogen 10-15%</td><td></td><td></td></tr>
<tr><td colspan="2">Secondary Elements ppm</td><td></td><td></td></tr>
<tr><td colspan="2">Boron B</td><td></td><td></td></tr>
<tr><td colspan="2">Iron Fe</td><td></td><td></td></tr>
<tr><td colspan="2">Manganese Mn</td><td></td><td></td></tr>
<tr><td colspan="2">Copper Cu</td><td></td><td></td></tr>
<tr><td colspan="2">Zinc Zn</td><td></td><td></td></tr>
<tr><td colspan="2">Aluminum</td><td></td><td></td></tr>
</table>

Kelp Typical Analysis

Maxicrop Kelp Meal Nutritional Analysis:

It contains the following vitamins: A, B_1, B_2, B_{12}, C, D, E, K, Riboflavin, Choline, Carotene and Pantothene.

Mineral content:

		%			
Ag	Silver	.000004	Mg	Magnesium	.213000
Al	Aluminum	.193000	Mn	Manganese	.123500
Au	Gold	.000006	Mo	Molybdenum	.001592
B	Boron	.019400	N	Nitrogen	1.467000
Ba	Barium	.001276	Na	Sodium	4.180000
Be	Beryllium	Trace	Ni	Nickel	.003500
Bi	Bismuth	Trace	O	Oxygen	undeclared
Br	Bromine	trace	Os	Osmium	Trace
C	Carbon	undeclared	P	Phosphorus	.211000
Ca	Calcium	1.904000	Pb	Lead	.000014
Cb	Niobium	Trace	Pd	Palladium	Trace
Cd	Cadmium	Trace	Pl	Platinum	Trace
Ce	Cerium	Trace	Ra	Radium	Trace
Cl	Chlorine	3.680000	Rb	Rubidium	.000005
Co	Cobalt	.001227	Rh	Rhodium	Trace
Cr	Chromium	Trace	S	Sulfur	1.564200
Cs	Cesium	Trace	Se	Selenium	.000043
Cu	Copper	.000635	Sb	Antimony	.000142
F	Fluorine	.032650	Si	Silicon	.164200
Fe	Iron	.089560	Sn	Tin	.000006
Ga	Gallium	Trace	Sr	Strontium	.074876
Ge	Germanium	.000005	Te	Tellurium	Trace
H	Hydrogen	undeclared	Th	Thorium	Trace
Hg	Mercury	.000190	Ti	Titanium	.000012
I	Iodine	.062400	Tl	Thallium	.000293
Id	Indium	Trace	U	Uranium	.000004
Ir	Iridium	Trace	V	Vanadium	.000531
K	Potassium	1.280000	W	Tungsten	.000033
La	Lanthanum	.000019	Zn	Zinc	.003516
Li	Lithium	.000007	Zr	Zirconium	Trace

Florida Colloidal Clay Phosphate Analysis

Below is a typical analysis for Lonfosco colloidal clay phosphate from Florida USA. For many years Lonfosco was the most popular and available sof rock phosphate. It is no longer readily available , but has been largely replaced by Calphos, which comes from the same area. As we have been unable to locate a typical analysis for Calphos, we are reproducing the Lonfosco table.

AVERAGE ANALYSIS OF "LONFOSCO" DRY BASIS	
Phosphoric Acid, P_2O_5	18.61
Iron Oxide (Fe203)	3.73
Aluminum Oxide (A1203)	15.47
Calcium Oxide (CaO)	22.17
Magnesium Oxide (MgO)	1.16
Sodium Oxide (Na2O)	.19
Potassium Oxide (K2O)	.52
Acid Insolubles	29.68
Silicon Dioxide (SiO2)	26.72
Sulfate Sulfur (S)	< 0.1
Carbon Dioxide (CO2)	2.02
Flouride (F)	1.65
Chloride (Cl)	.011
Boron (B)	< 100 ppm
Arsenic (As)	7.2 ppm
Cadmium (Cd)	1.6 ppm
Chromium (Cr)	63 ppm
Cobalt (Co)	< 5 ppm
Copper (Cu)	821 ppm
Lead (Pb)	187 ppm
Manganese (Mn)	72 ppm
Mercury (Hg)	3.5 ppm
Molybdenum (Mo)	45 ppm
Nickel (Ni)	38 ppm
Selenium (Se)	< .25 ppm
Vanadium (V)	16 ppm
Zinc (Zn)	< 1370 ppm

Annual Crop Uptake of Major Nutrients K_2O, Mg, and S

Annual Crop Uptake of the Major Nutrients K_2O, Mg, and S

How much sulfur, potassium and magnesium do typical crops use?
The following table gives the amounts of K2O, Mg and S used in the above-ground portion of various crops for a particular yield level. This information can serve as a useful guide in determining the crop's requirement for these three nutrients.

Nutrient Utilization (kg/Ha)

Crop	Yield	K2O	Mg	S
Cotton	1700	171	39	45
Corn	11300	269	56	34
Grain Sorghum	9000	207	49	43
Peanuts	4500	207	28	24
Rice	7800	166	16	13
Soybeans	2700	109	20	19
Alfalfa	18000	538	45	46
Coastal Bermudagrass	11000	470	56	45
Orchardgrass	13000	420	28	39
Fescue	8000	207	15	22
Sugar Beets	67000	616	90	50
Apples	13000	202	27	*
Peaches	37000	134	25	*
Oranges	27500	370	43	31
Tomatoes	30500	376	31	46
Tobacco, Burley	4500	362	36	46
Tobacco, Flue	3400	288	27	21
Wheat	5400	181	27	22
Potatoes, White	25000	612	56	25

* No Data

Chart courtesy of Sul-Po-Mag
www.sul-po-mag.com

pH and Nutrient Availability

pH and its relation to nutrient availability in high organic matter soils and mineral soils.

Note that in both types of soil the highest overall availability for the widest range of elements is between pH 6.0 and pH 6.5

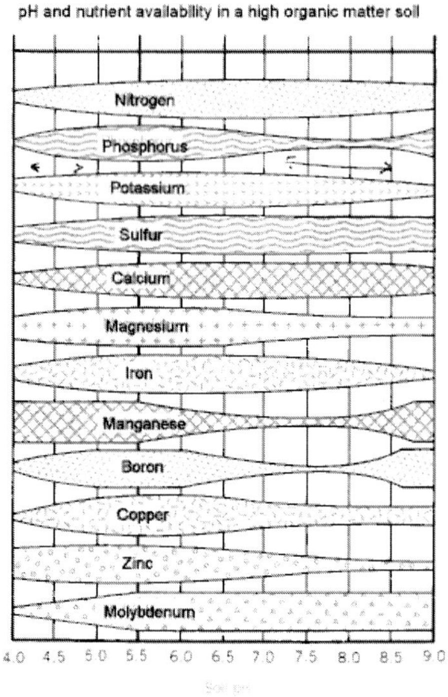

FIGURE 3.7 The relation between soil pH and the relative availability of plant nutrients in organic soils or Histosols. The wider the bar, the greater the availability. (From Lucas, 1982.)

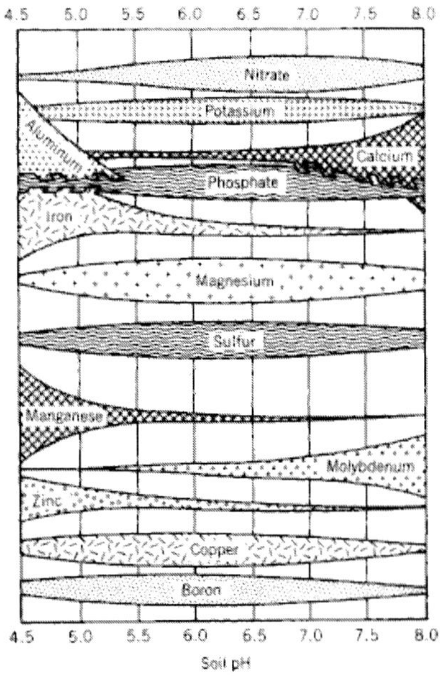

Note that this chart is showing the pH to nutrient availability for a *mineral* soil.

FIGURE 3.5 The relation between soil pH and the relative availability of plant nutrients in mineral soils. The wider the bar, the greater the availability. (From *Soil Handbook*, University of Kentucky, 1970.)

Reams' recipe for an ideal soil

Here is Carey Reams' recipe for a balanced soil as measured by the LaMotte soil test using the Morgan extractant

Calcium	2,000-4,000 lbs.
Magnesium	285-570 lbs.
Phosphate	400 lbs.
Potash	200 lbs.
Nitrate Nitrogen	40 lbs.
Ammonium Nitrogen	40 lbs.
Sulfate	200 lbs.
Sodium	20-70 ppm.
ERGS	200-600 micromohs
pH	6-7

Adapted from Arden Andersen Science in Agriculture
ERGS stands for Energy Release per Gram of Soil

Reams vs Albrecht? Why not both?

One bit of contention between the Reams and Albrecht schools of mineral balanced agriculture is that two different testing methods are used. The Albrecht school uses the standard soil test, also called the Brookside test, which employs strong acids and bases to extract and measure the amount of potentially available nutrient minerals. The Reams school prefers the Lamotte soil test, which uses a weaker extraction method called the Morgan solution. The Morgan extractant is closer to the pH and concentration of the acids that plants are able to produce, and shows the amounts of easily available mineral nutrients. An advantage of the LaMotte test is that it requires a minimum of training and equipment (less than $1000 in total cost) and can be done at home. In this author's opinion, both testing methods are equally valuable and have their place.

The standard soil test allows one to measure the exchange capacity and percentage of cation saturation in the soil, so that it can be balanced chemically, physically, and ionically. That balance is the basic premise and method of this book, The Ideal Soil. Once that has been done, the LaMotte test can be used as often as desired to check the day-by-day availability of the major nutrients during the growing season. The two tests are not interchangeable but they are very much complementary.

Some further confusion between the viewpoints of the Reams and Albrecht schools arises because of the apparent discrepancy between Wm Albrecht's recommendation of a Calcium to Magnesium ratio of 65% to 15%, versus Carey Reams' call for seven times as much Calcium as Magnesium. The following explanation will show that they are both saying the same thing:

Note the Reams numbers in the table above. Ca:Mg ratio is 2000 / 285, or exactly 7:1.

Albrecht's recommendation was for EC saturation: 65%Ca, 15%Mg

65 / 15 = 4.33:1

Why is Reams calling for 7:1 while Albrecht is saying 4.33:1? Because:

When it comes to ability to saturate exchange sites, Mg has about1.6 times the neutralizing ability of Ca, **by weight**. This is because Calcium has an atomic weight of 40, while Magnesium's atomic weight is only 24. Magnesium is a smaller and lighter atom, but both Ca and Mg have a ++ charge and each has the ability to fill two negative - exchange sites.
In the Reams chart above, if the soil contains 2,000lbs of Calcium, the recommendation is for 285lbs of available Magnesium, and this Magnesium has, as noted, about 1.6 times as much exchange site filling ability as an equal weight of Calcium.

285lbs (Mg) x 1.6 = 456 lbs

In other words, 285 lbs of Mg has the same acid-neutralizing or exchange capacity filling ability as 456 lbs of Ca.

If we redo the earlier division using the new number we get:

2000 / 456 = 4.39:1 or almost the same ratio (4.33:1) that Albrecht called for. Reams is talking weight, Albrecht is talking EC saturation. They are both calling for the same ratio of Ca to Mg..

Pounds, Acres, Kilograms and Hectares:

The convention used for estimating lbs/Acre in the English/Avoirdupois system is that the top 7" (17.8cm) of an acre of soil weighs 2,000,000 (two million) pounds, so one part per million (1 ppm) = 2 lbs/acre.

The convention used for estimating kilograms per hectare (kg/ha) is that the top 15 cm (5.9") of a hectare of soil weighs 2,000,000 kg, so 1 ppm = 2 kg/ha.

Considering the huge variance in soil densities, from light weight peat-type soils to heavy clays, unless one wishes to dig up, dry, measure, and weigh a volume sample of the particular soil they are working with, it's safe enough for agricultural purposes to simply say:

1ppm = 2 lb/acre = 2 kg/hectare

1ppm = 20g/1000ft^2 = 20g/100m^2

Conversion Factors:

The top 6 to 7 inches of 1 acre of average soil is assumed to weight 2 000 000 lbs
1 acre = 43 560 ft^2 (To convert lbs/acre to lbs/1000 ft^2, divide by 43.56 or 44)
The top 15 to 17 cm of 1 hectare of average soil is assumed to weigh 2 000 000kg

1 hectare = 10 000 meters^2

1 acre = 0.405 hectare

1 hectare = 2.47 acres

100 meters^2 = 1078 ft^2

1 part per million (ppm) = 2 lbs/acre or 2 kg/ha

1 lb per acre = 1 gram per 100 ft^2

1 kg per hectare = 1 gram per 10 m^2

1 ppm = 2 grams per 100 ft^2 or 2 grams per 10 m^2

1 ppm = 20 grams per 1000 ft^2 or 20 grams per 100 m^2

1 kilogram = 1000 grams

1 pound = 0.454 kg (454 grams)

1 kilogram = 2.2 lbs

Copper

Copper (Cu) is element number 29 on the Mendeleyev chart, the Periodic Table of the Elements. The other elements in Copper's specific group (group 1B, directly below it on the table) are Silver (Ag) and Gold (Au), which puts it in some racy company.

Copper is the key to elasticity in the plant. It is an important constituent of many proteins like ascorbic acid oxidase, cytochrome oxidase, diamine oxidase, and polyphenol oxidase. Copper is an important nutrient for many microbes, such as Aspergillus niger. It controls molds and often alleviates perceived zinc deficiencies. Copper interacts with iron and manganese. Andersen *Science In Agriculture* p236

Bordeaux mixture and Burgundy mixture are two famous sprays used to control fungus in vineyards. Developed in their eponymous provinces of France, Bordeaux mix is copper sulfate, mason's lime (calcium hydroxide), and water; Burgundy mix is copper sulfate, sodium carbonate (washing soda), and water. The full recipes and instructions for using Bordeaux and Burgundy mixtures are given below in the section borrowed from the Copper Development Association's web site.

The story goes that Bordeaux mixture was discovered by accident. During a wet fall in the province of Bordeaux in the 1880s the grapes were being severely attacked by downy mildew. Along a road that ran past one vineyard, the owners had sprayed a mixture of copper and lime on the vines, which turned the grapes a blue green color and was meant to dissuade the passers by from picking the grapes. The French scientist Millardet, while walking along, noticed that those vines were not being attacked by the fungus, and Bordeaux mixture was born.

As a part of Bordeaux mixture in grape arbors, it functions as a nutrient and not as an insecticide as is often believed. Walters, *Eco-Farm* p136 [Copper's use in Bordeaux mixture is actually as a fungicide, not an insecticide, but we'll allow Charles Walters the occasional typo. This observation should actually be credited to William Albrecht, who theorized that the copper in the mixture was stimulating the plant's immune system.]

Copper, vitally important to root metabolism, helps form compounds and proteins, amino acids and a host of organic compounds. It acts as a catalyst or part of the enzyme systems. It helps produce dry matter through stimulation of growth, prevents development of chlorosis, rosetting and dieback. Walters *Eco-Farm* p 197

The role of organic matter in Cu chemistry is also indicated by analysis of the soil solution. More than 99% of the Cu in the soil solution is complexed by organic matter. This complexing is of great importance in maintaining adequate Cu in solution for plant use. Foth and Ellis *Soil Fertility* p141

Because Cu is not translocated in the plant, the deficiency symptoms appear on the new growth. In small grains and corn the leaves appear olive or yellowish green in color, and often the leaves fail to unroll as they emerge. Often the leaf tips will appear as though the plants have been frost-damaged, and there will be some flags. A *flag* is a wilted or dead leaf or a branch with such leaves on an otherwise healthy appearing plant. *Soil Fertility* p157

Sul-po-mag, [also known as K-Mag and Langbeinite] applied between July 15 and September 15 up to 200 lbs per acre, seems to help in copper availability. *Science in Agriculture* p236.

Copper in Human and Animal Health

An excess of copper results in degeneration of the liver. It causes blood in urine and poor utilization of nitrogen.
A deficiency of copper is created by excess of molybdenum and cobalt. It produces anemia due to poor iron utilization. It depresses growth. Other symptoms...depigmentation of hair and abnormal hair growth; impaired reproductive performance and heat failure; scouring, fragile bones; retained placenta and difficulty in calving; and muscular incoordination in young lambs, and stringy wool. Walters *Eco-Farm* p367

...a largely vegetarian diet lacks the fat-soluble catalysts needed for mineral absorption. Furthermore, phytates in grains block absorption of calcium, iron, zinc, copper and magnesium. Unless grains are properly prepared to neutralize phytates, the body may be unable to assimilate these minerals. Fallon and Enig *Nourishing Traditions* p27

Ragweed, for example, is generally indicative of a phosphate/potash imbalance, but, more specifically, it indicates a copper problem. Copper is important in the use of manganese and iron, as well as in many metabolic reactions, Copper also seems to be important in controlling fungal disorders. Many people have allergic reactions to ragweed pollen. This reaction seems to be related to a copper deficiency in the mucous membranes. Andersen *Science In Agriculture* P.192

Copper: Needed for the formation of bone, hemoglobin and red blood cells, copper also promotes healthy nerves, a healthy immune system and collagen formation. Copper works in balance with zinc and vitamin C. Along with manganese, magnesium and iodine, copper plays an important role in memory and brain function. Nuts, molasses and oats contain copper but liver is the best and most easily assimilated source. Copper deficiency is widespread in America. Animal experiments indicate that copper deficiency combined with high fructose consumption has particularly deleterious effects on infants and growing children. *Nourishing Traditions* p43

Many enzymes incorporate a single molecule of a trace mineral-- such as manganese, copper, iron or zinc-- without which the enzyme cannot function. *Nourishing Traditions* p46
Graeme Sait: Can you revert grey hair with copper supplements? I've had grey hair since I was twenty-five.
Joel Wallach: It's definitely a Copper deficiency, and you could revert to your former hair color if you addressed the problem. I see it every day with my clients. It can be quite humorous when a seventy year old grey-haired man returns to his former redheaded glory. Sait, *Nutrition Rules* p297

In Australia it was discovered that black sheep grazing on copper-deficient pastures turned gray.

In humans copper is stored in the liver. In cases of fever and infection, the level of iron in the bloodstream drops and the blood copper level rises as the copper reserves in the liver are mobilized to aid the immune system in fighting off invaders. This tidbit is from Andre

Voison's classic *Soil, Grass, and Cancer* , in which the French bio-chemist and veterinarian devoted several chapters to the role of copper in human and animal health.

In the 1930s Dr. Weston A. Price investigated the traditional diets of isolated peoples around the world. High in the Andes mountains of South America he discovered the native peoples relied on dried fish eggs and seaweed brought from the ocean to supply trace minerals and other factors lacking in their diet. He writes "The kelp provided a very rich source of iodine as well as copper, which is very important to them in the utilization of iron for building an exceptionally efficient quality of blood for carrying oxygen liberally at those high altitudes. W. A. Price, *Nutrition and Physical Degeneration* p 265

Copper functions in the body as an enzyme co-factor, formation of hemoglobin and red blood cells, protein metabolism, synthesis of phospholipids, vitamin C oxidation, production of elastin, and formation of RNA. Signs of possible deficiency are white hair, liver cirrhosis, allergies, parasites, hernia, anemia, hyper/hypo thyroidism, arthritis, ruptured disc and iron storage disease. Walters, *Minerals for the Genetic Code* p122.

Zinc and copper have a seesaw relationship in the body, competing with each other for absorption in the gut. Both zinc deficiency and copper toxicity have increased since the switch from zinc (galvanized) to copper water pipes. We can avoid this problem by not drinking tap water. Haas, *Staying Healthy with Nutrition* p191

The following wealth of information is from the **Copper Development Association** website:

Uses of Copper Compounds: Copper Sulphate's Role in Agriculture

Copper sulphate has many agricultural uses but the following are the more important ones:

- Preparation of Bordeaux and Burgundy mixtures on the farm
- Control of fungus diseases
- Correction of copper deficiency in soils
- Correction of copper deficiency in animals
- Stimulation of growth for fattening pigs and broiler chickens
- A molluscicide for the destruction of slugs and snails, particularly the snail host of the liver fluke

Preparation of Bordeaux and Burgundy Mixtures on the Farm
Because of their importance to farmers, instructions concerning the dissolving of copper sulphate and the preparation of both Bordeaux and Burgundy mixtures have been included in the text.
Dissolving Copper Sulphate
Iron or galvanized vessels must not be used for the preparation of copper sulphate solutions. Plastic vessels, now freely available, are light and very convenient. To make a strong solution, hang a jute sack of copper sulphate so that the bottom of it dips a few inches only in the water. The copper sulphate will dissolve overnight. Copper sulphate dissolves in cold water to the extent of about 3 kg per 10 litres. If more than this is placed in the sack described above, then a saturated solution will be obtained and it may be used without serious error on the basis that it contains 3 kg copper sulphate per 10 litres.

Preparation of Bordeaux Mixture

Bordeaux mixture is prepared in various strengths from copper sulphate, hydrated lime (calcium hydroxide) and water. The conventional method of describing its composition is to give the weight of copper sulphate, the weight of hydrated lime and the volume of water in that order. The percentage of the weight of copper sulphate to the weight of water employed determines the concentration of the Bordeaux mixture. Thus a 1% Bordeaux mixture, which is the normal, would have the formula 1 :1:100the first 1 representing 1 kg copper sulphate, the second representing 1 kg hydrated lime, and the 100 representing 100 litres (100 kg) water. As copper sulphate contains 25% copper metal, the copper content of a 1% Bordeaux mixture would be 0-25 % copper. The quantity of lime used can be reduced considerably. Actually 1 kg copper sulphate requires only 0.225 kg of chemically pure hydrated lime to precipitate all the copper. Good proprietary brands of hydrated lime are now freely available but, as even these deteriorate on storage, it is safest not to exceed a ratio of 2:1. i.e. a 1:0.5:100 mixture.
In preparing Bordeaux mixture, the copper sulphate is dissolved in half the required amount of water in a wooden or plastic vessel. The hydrated lime is mixed with the balance of the water in another vessel. The two "solutions" are then poured together through a strainer into a third vessel or spray tank.
[Note from soilminerals.com: It appears from the above that one doesn't want to mix the lime solution with the copper solution until one is ready to spray, as the lime precipitates the copper. Well, that makes sense; copper sulfate is acid, calcium hydroxide is alkaline, mix them together and what happens? You probably get calcium sulfate (gypsum) and copper oxide. Copper oxide is not water soluble.]

Preparation of a 1% Burgundy Mixture
Dissolve separately 1 kg copper sulphate in 50 litres water and 1.25 kg sodium carbonate (washing soda) in 50 litres water and slowly add the soda solution to the copper sulphate solution with stirring. Control of fungus diseases
Bordeaux and Burgundy mixtures have been found effective in controlling a whole host of fungus diseases of plants. Normally a 0.5 % to 1 % Bordeaux or Burgundy mixture applied at 2 to 3 week intervals suffices to control most copper-susceptible fungi.
Generally once the fungus spores have alighted on the host plant and penetrated the tissues it is difficult to control them. The principle of control must in most cases depend on protection, i.e. preventing the fungus spores from entering the host tissues. Copper fungicides are noted for their tenacity and for this reason are much to be preferred in areas of high rainfall.

The simplest method of control is to apply a protective coating of Bordeaux or Burgundy mixture (or other copper fungicide) to the susceptible parts of the plant, so that spores alighting on them come in contact with the protective film of copper and are killed instantly. It is thus important to remember that the first spraying must ideally be made just before the disease is expected and continued at intervals throughout the susceptible period. For this reason it is important to take advantage of the early warning schemes which are in operation to ensure greater accuracy of the timing of the first spraying.
It must also be remembered that fungi are plants and that control measures that will kill them may not always leave the host plant unaffected. The use of too concentrated a fungicide mixture must therefore be guarded against, particularly for the early sprays.
Copper fungicides have been reported effective against numerous plant diseases. A list, by no

means exhaustive, of some 300 diseases that have been found amenable to control by copper fungicides. [**note: the list is at the CDA web site**]

Correction of Copper Deficiency in Soils
Where copper deficiency has been confirmed by soil analysis or field diagnosis, whether in plants or animals, it can be corrected very simply either by applying 50 kg copper sulphate per hectare in the form of a fertilizer before sowing or by spraying the foliage of the young cereal plants, when they are about 150 mm high, with 750 grams copper sulphate (dissolved in from 400 to 2,000 litres water) per hectare. The soil application has generally given the better results and has the advantage that it may have a residual effect for more than ten years. The foliar application has to be given annually to each crop. An alternative is to add a copper containing slag (normally about 1% to 2 % copper) at a rate of a tonne to the hectare.

[Note from soilminerals.com: 50kg per hectare of copper sulfate works out to about 5-6 ppm of elemental copper]

Correction of Copper Deficiency in Animals
A method of correcting copper deficiency in livestock is to treat the soil on which animals graze. For example, in Australia and New Zealand swayback in lambs is being prevented by top dressing copper deficient pastures with 5 to 10 kg copper sulphate per hectare some time before lambing begins.
Other methods include drenching periodically with a copper sulphate solution; incorporating copper sulphate in salt and other animal licks; or by what is probably the most general method, incorporating copper sulphate along with other minerals and vitamins in the form of carefully blended supplements in the feeding stuffs.

Stimulation of Growth for Fattening Pigs and Broiler Chickens

The inclusion of up to as much as 0.1% copper sulphate in the diet of bacon and pork pigs and broiler chickens stimulates appetite and produces increased growth rate with a marked improvement in feed conversion.
A molluscicide for the destruction of slugs and snails, particularly the snail host of the liver fluke. All likely habitats of the liver fluke snail should be treated with copper sulphate at the rate of 25 kg to the hectare at least twice a year in June and August (northern hemisphere) or December and February (southern hemisphere).

End of info from CDA web site

At SoilMinerals.com we use and recommend Copper in the form of Copper sulfate, containing 25% Cu. It is highly water soluble for use as a soil amendment, a foliar spray, a fertilizer ingredient, or for making Bordeaux or Burgundy mixtures as described above.

Copper sulfate is available as an animal feed supplement; this is often the best source when large quantities are needed.

Zinc

First we hear from Arden Andersen, who thinks zinc is overused in agriculture:

Zinc is an essential component of many enzymes in the dehydrogenase, proteinase, and peptidase groups. It is a minor catalyst for sul-po-mag and copper and is correlated closely with copper and active nutrient systems. Zinc helps to make acetic acid in the root to prevent rotting; it is used to control blight and allows dead twigs on trees to shed off. Perceived zinc deficiency is often only symptomatic. Research has indicated that known soil-zinc deficiencies result in symptoms of plant-zinc deficiency only about 50% of the time. Zinc is much overused and promotes the growth of many weed species. Andersen *Science in Agriculture* p238

And next from Gary Zimmer, who appears to be a big fan of Zinc, particularly for corn/maize:

Zinc-- contributes to test weight, increased corn ear size, promotes corn silking, hastens maturity, chlorophyll formation, enzyme functions, regulates plant growth. Zimmer *The Biological Farmer* p109 [Zimmer also writes that zinc is "essential for corn starters" and recommends 5 lbs/acre of 35% zinc to supply a corn crop and build soil levels.]

And a few more experts weigh in on Zinc:

Charles Walters says that Zinc "may act in the formation of chlorophyll.[....]It certainly stimulates plant growth and prevents the occurrence of mottled leaf in citrus, white bud in corn, and other disorders." He further states that "Plants do require it in the 3-100ppm range." and regarding animal health that "An excess of zinc means decreased copper availability and interference with utilization of copper and iron, bringing about anemia. A zinc excess also shows up as bald patches and skin disorders (rough skin), a deficiency is created by excess of calcium. Zinc is absolutely essential for production of sperm. It also increases the need for vitamin A." Walters *Eco-Farm* p366.

Now a word from the more mainstream guys: " ...zinc uptake by plants declines as pH increases.[....] High levels of phosphorus in soils has been known to intensify zinc deficiency in a number of crops. The exact cause of the zinc-phosphorus antagonism has been difficult to determine....the zinc-phosphorus antagonism occurs on calcareous [high calcium] soils and may be related to iron availability." Foth and Ellis *Soil Fertility* p142 They also show an increase from 4.2 to 19.9 bushels per acre of pea beans on one field after the addition of 25lbs/acre of zinc, quite the boost.

Here's a fun one from an interview with Klaas Martens in Graeme Sait's Nutrition rules. "...we need to lift our zinc levels as our phosphorus levels increase. We always need to use zinc with our starter fertilizers. At one time, our consultant suggested that we had a zinc deficiency, simply by driving past one of our fields. He didn't need a soil test, because the presence of milkweed was an indicator of a zinc shortage. We've actually seen the milkweed disappear as we have slowly corrected the zinc."

The area of South-central Washington state known as the Palouse is one of the world's great wheat growing regions. When it was first broken to the plow the production was tremendous, but by the 1920s it had fallen dramatically. The problem turned out to be zinc insufficiency. Zinc is easily water soluble, and this fact combined with low initial reserves of zinc in many soils has made zinc deficiency common. It was also one of the earliest trace mineral deficiencies discovered, and its sometimes dramatic effect on crop yields has led to some overuse; one book in front of me lists the results from application of 122 lbs/acre of zinc! The results were quite disappointing, understandably.

Both zinc and copper are well known for their need in animal nutrition, and most commercial livestock producers supplement animal feed with these minerals. For that reason, manures from commercial livestock operations are frequently very good sources of zinc and copper (and sometimes boron). The problem with these manure sources of minerals is that one doesn't know how much they are getting, or what else they may be getting that they don't want.

Moving on to the human nutrition aspect, the adult human body contains about 2400 milligrams of zinc. Zinc is most concentrated in the male prostate and semen. The next most concentrated tissues are the retina of the eye, the heart, spleen, lungs, brain, and adrenal glands. Because of zinc's role in RNA and DNA synthesis and in the formation of many enzymes, zinc deficiency leads to slow healing of wounds. In some hospital tests zinc supplements led to surgical incisions healing in one-half the "normal" time. Zinc is important to normal insulin activity, the functions of taste and smell, normal immune function, protein digestion, and the formation of bones and teeth as it is a co-factor of alkaline phosphatase. Fallon and Enig, in *Nourishing Traditions* , call zinc the "intelligence mineral". It is generally more easily absorbed from animal products than from plants and although grains may contain significant zinc, that zinc may be bound up by the phytates in the grain's outer portion. Many traditional peoples soaked and sprouted seeds and grains before cooking them, a practice that reduces or eliminates this mineral-binding by phytates. Fallon and Enig add "Even a minor zinc deficiency in pregnant animals results in offspring with deformities, such as club feet, cleft palates, domed skulls and fused and missing ribs. In humans, zinc deficiency can cause learning disabilities and mental retardation."

Some of the other human nutritional and health problems associated with zinc deficiency are acne, boils, psoriasis, gastric ulcers (zinc is needed to form digestive acids), cataracts, hypertension, infertility, loss of or poor functioning of the senses of hearing, taste, and smell, weak muscles, and fatigue.

The brilliant British researcher Mark Purdey, in his groundbreaking work with mad cow disease and chronic wasting disease, found in a worldwide survey that both mad cow and CWD were strongly associated with soils that had very low levels of zinc and copper, combined with high levels of manganese and sometimes high levels of strontium and silver. In those conditions copper in the melanin granules, which are transmitters of outside information to the brain, may be replaced by manganese with disastrous results. In other words, neither mad cow disease nor chronic wasting disease are caused by infectious microbes, but are the result of a mineral imbalance.

Purified Source	Sulfur S	Zinc Zn
Zinc sulfate 1H_2O	**17**	**35**
Zinc sulfate 7H_20**	**11**	**22**

****Highly soluble in H_2O**

Iron

Iron is second only to aluminum in the list of abundant metals. It makes up about 5% of the earth's crust, so it is rarely absent from soils, although it may not be present in an available form.

For garden soil we like to see 50-200ppm of iron on a standard soil test. Above 250 ppm may indicate something out of balance.

What does iron do in the plant? Paraphrasing Arden Andersen, "Iron draws energy to the leaf by absorbing heat from the sun; it makes the leaf darker, thus absorbing more energy. It will increase the waxy sheen of the crop. Iron is necessary for the maintenance and synthesis of chlorophyll and RNA metabolism in the chloroplasts. It increases the thickness of the leaf, [which] increases nutrient flow geometrically, resulting in a production increase geometrically." *Science in Agriculture* p236

Iron is needed by nitrogen fixing bacteria.

So iron is a good thing, in most cases. Below we have a couple of different views on just how good it is and how much we want:

Both iron and manganese become less available at pH 7 and above and in the absence of organic matter and water. These conditions are found in some arid parts of the western United States. High calcium soils also tend to have low available iron, particularly if they are also low in organic matter. In a calcareous soil, most of the potentially available iron is tightly bound to organic matter. Some plant roots have been shown to have the ability to obtain iron from these sources by chemically reducing ferric iron (Fe+++) to ferrous iron (Fe++). High phosphorus soils may also have low available iron, as any free iron will chemically bind to from iron phosphate....Correcting an iron deficiency may be difficult because the problem is not a lack of iron in the soil, but that it is chemically bound. Lowering the pH, if practical, is the surest method. Foliar iron sprays are also effective. Foth and Ellis *Soil Fertility* pp146-147

Here's an excerpt from an interview with Gary Zimmer "In our dairy work we are looking at phosphorus as a key element. We want phosphate uptake for sugars and energy and digestibility and plant health. If I have high iron in my soils, usually from over-tillage, excessive use of caustic materials or too much nitrogen use, I'm not happy. On a dairy farm, I scream and holler if they buy a single pound of commercial nitrogen. If they buy nitrogen, I want to know why. They had better use their manures and alfalfa and rotation, because I don't want iron buildups. **Iron binds with phosphate within the plant** . Many people who don't feed cattle don't notice this difference. You see, the phosphorus may be in the plant, but when you bind it to iron, it becomes unavailable. Iron has a triple-positive charge and phosphorus has a triple-negative charge, so they will bond very easily. If your feed is high in iron, then the cow is starved for phosphorus. **We are fanatical about trying to get our iron down, just so we have better phosphate availability** . In high iron soils I don't think our soil tests give an accurate idea of phosphate availability to the plant." [emphasis added] Graeme Sait *Nutrition Rules* pp187-188. Gary Zimmer works mostly with neutral or alkaline pH soils in the upper Midwestern US, and we don't know offhand what he considers high iron.

The info above brings up some interesting questions about iron supplements in general, don't you think? I wonder what connection there might be between the high iron intake recommended for women and high incidences of osteoporosis? Fallon and Enig have this to say about one type of iron supplementation "Recently, researchers have warned against inorganic iron used to supplement white flour. In this form, iron cannot be utilized by the body and its buildup in the blood and tissues is essentially a buildup of toxins. Elevated levels or inorganic iron have been linked to heart disease and cancer." *Nourishing Traditions* p44.

Charles Walters has this to say about signs of iron deficiency in plants "When iron deficiency is serious, the entire leaf will turn yellow, leaving only the veins to stand out like road maps....Chlorosis (white leaves that should be green) is possible even in the presence of iron. Lime can complex iron, and yet in the human being calcium and copper must be present for iron to function properly. In order to free iron, the farmer must complex calcium in this case, and this means using either iron sulfates or iron chelates, or substituting a proper foliar blend." *Eco-Farm* p196.

At soilminerals.com we have seldom seen a soil test that showed a lack of iron. and as we often are working with gardens and fields of a few acres and smaller, our approach is to bring the pH down below 7 which will make iron (as well as the other cations) more easily available. The alternative, if one cannot lower the pH with minerals because of size, expense, highly calcareous soils, or other constraints, is to increase the biological activity in the soil. As noted above in the excerpt from Foth and Ellis' Soil Fertility, in a calcareous soil most of the iron is tied up with organic matter. Increasing the organic matter content of such soils will provide more holding points for iron, and increasing the biological activity, through the addition or seeding of beneficial bacteria and fungi, should make more Fe available to the plants.

Jersey Greensand (Glauconite) is also a good source of Iron, averaging around 7% Fe. Many sources of rock dust (crusher dust) are also high in Iron.

Purified Source	Sulfur S	Iron Fe
Fe sulfate 1H_2O	**18**	**30**
Fe sulfate 7H_20**	**11.5**	**20**

****Highly soluble in H_2O**

MANGANESE

Manganese is synergistic with iron; they work together in biology in ways that are not well understood, but we do know that they need each other. Good steel must have some manganese in it to impart toughness, and that manganese in the steel also absorbs oxygen during the steel making process. Perhaps this is a clue to the biological relationships of Mn and Fe, in that the manganese may slow the oxidation rate of iron in living things.

We at SoilMinerals.com like to see about 1 part manganese to 2 parts iron on soil test results, up to about 50 ppm manganese. Levels above 50 ppm may be too high, particularly if the soil is deficient in copper and zinc. In wet, acid soils below pH 5 or so that naturally contain high amounts of manganese, soluble manganese can reach levels that are toxic to plant roots. The remedy for these conditions would be to drain the soil better, or, if the crop requires a wet, acid soil (e.g. cranberries), the remedy would be to increase water flow through the soil, as more water will bring more oxygen, which will precipitate the excess manganese in an insoluble state.

Arden Andersen calls manganese "the element of life", and says that manganese "brings the electrical charge into the seed, creating the magnetic force to draw the other elements into the seed." *(Science in Agriculture* p236.) In *Eco-Farm*, Walters credits manganese with aiding the oxidase enzyme in carrying oxygen, and entering into the oxidation and reduction reactions needed in carbohydrate metabolism and in seed formation; more clues that manganese has a strong connection with oxygen. Regarding manganese in animal nutrition, Walters tells us that an excess of manganese increases the need for iron, while a manganese deficiency results in leg deformities in calves, eggs not formed correctly, degeneration of testicles, offspring born dead, and delayed heat periods, and also says that an excess of calcium and phosphorus may lead to a manganese deficiency. *(Eco-Farm* p366)

We definitely know that manganese is necessary for the development of viable seeds. The most common and obvious sign of manganese deficiency is in the almond family. Peaches, nectarines and apricots with split-open pits containing a shriveled seed are the prime example. Dan Skow has some interesting insights on this from the Carey Reams school of thought: "If there is no Manganese in the seed, it will swell up and rot [rather than sprouting]. Manganese has a high atomic weight, 54.9380, meaning it has more power than nutrients in the surrounding soil. [Manganese] puts into play the magnetism necessary to draw nutrients into the seed to feed it and its emerging root system. When there is a shortfall for manganese, the entire fertility program has to be adjusted to create enough energy to pull more manganese." (*Mainline Farming for Century* 21p59.) Skow recommends a foliar spray of manganese mixed with phosphoric acid to easily correct manganese deficiency problems, and tells us that manganese is what is needed to ensure regular pecan crops with filled hulls.

Moving on to human nutrition, Elson Haas tells us that manganese is an essential part of the superoxide dismutase enzyme found in the mitochondria, the energy factories in the cells. Manganese also activates the enzymes necessary for the body to use biotin, thiamine (B 1), vitamin C, and choline. *(Staying Healthy with Nutrition* p207). Sally Fallon writes that manganese is "..needed for healthy nerves, a healthy immune system and blood sugar regulation....also plays a part in the formation of mother's milk and in the growth of healthy bones. Deficiency may lead to trembling hands, seizures, and lack of coordination. Excessive milk consumption may cause manganese deficiency as calcium can interfere with manganese absorption...phosphorus antagonizes manganese as well. *(Nourishing Traditions* p44).

Manganese can also be quite toxic. It has been (likely still is) used as a flux or anti-oxidant coating on arc-welding rods, and some long-time welders have ended up with chronic and acute symptoms much like those listed above for manganese deficiency: trembling hands and other indications that appear identical to Parkinson's disease.

Manganese, we see, as well as being necessary, can be toxic, especially in diets or soils that are deficient in copper, zinc, and perhaps iron. The paragraph below was already posted above under zinc, but bears repeating:

The brilliant British researcher Mark Purdey, in his groundbreaking work with mad cow disease and chronic wasting disease, found in a worldwide survey that both mad cow and CWD were strongly associated with soils that had very low levels of zinc and copper, combined with high levels of manganese and sometimes high levels of strontium and silver. In those conditions copper in the melanin granules, which are transmitters of outside information to the brain, may be replaced by manganese with disastrous results. In other words, neither mad cow disease nor chronic wasting disease are caused by infectious microbes, but are the result of a mineral imbalance.

Purified Source	Sulfur S	Mang. Mn
Mn sulfate 1H_2O*	**19**	**32**

***Varies in solubility in H_2O**

BORON

Boron is one of the rarest elements, and one of the most mysterious. It is absolutely essential for calcium metabolism, but no one seems to know its method of action. An often heard phrase in the eco-agriculture field is "Calcium is the truck, but boron is the driver". This refers to the concept that calcium is the transporter of nutrients into and out of the cells, but it can't do its job unless boron is present.

There are apparently only two commercially viable boron deposits in the world, one in Turkey and one in the Mojave desert of Southern California. (Note: Since writing this, the author has learned that Chilean nitrate is also a good source of Boron. See the chapter on Minor Minerals in this book for more info on Chilean nitrate.)

Boron is easily leached out of soils, so higher rainfall areas are often deficient. In front of me is a map of the USA showing boron deficiency areas. Essentially it shows everything east of the Mississippi River as boron deficient, as well as the Pacific NW as far south as the San Francisco Bay and as far east as central Montana.

Here's Charles Walters on boron: "Plants must have boron, again in the trace range. Texts quote 2 to 75 parts per million as being essential, but note that plants vary in their required amounts according to species. Boron is quite lethal to seeds when used in the salt form." (*Eco-Farm* p136). 2 to 75 parts per million is a huge range. At soilminerals.com we would be very concerned to see available boron above 5ppm. Our general rule is 1 part of boron to 1000 parts calcium.

More on boron from Walters' *Eco-Farm* : "Boron is required so that calcium can perform its metabolic chore. It is essential in several other metabolic processes...it prevents such abnormalities as cracked stem in celery, internal cork in apples, black heart in beets and turnips, yellowing of alfalfa leaves. When boron deficiency is a problem, death of the terminal bud is a common symptom. Lateral buds continue to produce side shoots, but terminal buds on these side shoots fade away. Rebranching may occur, but the multi-branched plant will take on the appearance of a rosette.
In cauliflower, heads fail to mature properly and remain small. Reddish-brown areas become evident. Terminal buds take on a light green color...root crops are affected by brown heart, dark spots, or by splintering and cracking at the middle in...spuds [potatoes], sweet potatoes, radishes, carrots.
Boron is required for translocation of sugar, and this means boron deficiency can be spotted as a sugar deficiency. Important as it is, a 100 bushel crop of corn requires only 4 ounces of boron...a ton of alfalfa requires only a single ounce...boron regulates flowering and fruiting, cell division, salt absorption, hormone movement and pollen germination, carbohydrate metabolism, water use, and nitrogen assimilation.
In most soils boron is [found] as highly insoluble tourmaline, the supply being somewhere between 20 and 200 pounds per acre. It takes life in the soil to draw on this bank account, and the Creator has supplied this life in the form of microorganism species which simply have to have boron to live. By using the nutrient themselves and then contributing their bodies to the soil's fertility load, microorganisms change boron into an organic form.
When dry weather hits, microorganisms in soil without tilth and structure go dormant. This means the boron supply is cut off. Generally speaking there is more boron in the subsoil...and roots...dig deeper...for both moisture and for this very essential nutrient.
Too much boron will...restrict growth, cause plants to exhibit that sickly pale green color

sometimes mistaken for nitrogen deficiency, preside over root deterioration and poor yield. In short, either a shortage or marked imbalance of boron will set up a plant for insect and fungal attack."

Important stuff, Boron. It also has several more esoteric uses and connections, such as remediation of radiation poisoning. According to another Charles Walters book, *Minerals for the Genetic Code* (based on the work of Dr. Richard Olree), boron controls all the +3 charges in the human body, and it is easily displaced by aluminum, losing three boron molecules to every one aluminum molecule. Furthermore "Boron has the ability to absorb radiation and release it without changing the neutron. The heart is the most active part of the body for which reason boron defends the heart. The story has been told that Soviet truck drivers were offered bonuses to deliver boron to the Chernobyl site, this with the knowledge that their trip would be fatal, but families would be paid. None realized that, fortified with boron [themselves], they could have made their decision with impunity. Boron stopped the "China Syndrome" from occurring in Russia." [ed. note: as is often the case, Walters is being a bit obscure here. He appears to be stating that large quantities of boron were dumped on the nuclear pile at Chernobyl to stop the out-of-control nuclear reaction, and that if the truck drivers had swallowed some of that boron they would have been protected from radiation exposure.]

Continuing the quote on boron from *Minerals for the Genetic Code* : "Boron is known as the calcium helper and for the metabolism of calcium, magnesium and phosphorus. Boron improves retention of both calcium and magnesium and elevates circulation of serum concentrations of testosterone.

"Boron works in the body toward brain function, activates vitamin D, promotes electrical brain activity, enhances memory, and promotes alertness. Signs of possible deficiency include ADD/ADHD, osteoporosis, arthritis, fatigue, decreased motor function, decreased short-term memory, decreased brain function, and increased loss of calcium and magnesium in the urine."

As if all that wasn't enough, boron in the form of boric acid is our safest and most effective ant control, and is used in many areas to treat wood in ground contact from ant and termite damage, as well as being used to fire-proof cellulose insulation and as a flux for soldering and brazing metal. 20 Mule Team Borax, available in the laundry soap section of most grocery stores, is a pure and natural mined product containing about 10% boron. It is very suitable for garden use in small quantities. 7 ounces of 20 Mule Team Borax per 1000 square feet equals approximately 1 part per million of boron. Take it easy. As noted above, a boron deficiency can be induced simply by dry soil. Don't add boron without a soil test that indicates a need for it. 1-2 ppm per year is the maximum we recommend.

At SoilMinerals.com we use either Solubor, a concentrated Sodium borate that is 20% B, or regular 20 Mule Team borax, such as is used for laundry. 20 Mule Team borax is 9% B. Both are water soluble and easily used for soil applications, fertilizer mixes, or foliar feeding.

Source	Boron B %
Borax**	**9**
Solubor™**	**20.5**

****Highly soluble in H_2O**

References Cited and/or Used for the above essays on Copper, Zinc, Iron, Manganese, and Boron:
(In No Particular Order)

Eco-Farm by Charles Walters and C. J. Fenzau *Acres USA 1996*
Soil Chemistry 2nd Edition by Bohn, McNeal, O'Connor *Wiley-Interscience 1985*
Science in Agriculture by Arden Andersen *Acres USA 2000*
Mainline Farming for Century 21 by Skow and Walters *Acres USA 1995*
Staying Healthy with Nutrition by Elson Haas *Celestial Arts 1992*
Nutrition and Physical Degeneration by Weston A. Price *Price-Pottenger Nutrition Foundation 1939/2004*
Biological Farmer, the by Gary F. Zimmer *Acres USA 2000*
Soil Fertility by Foth and Ellis *John Wiley and Sons 1988*
Nutrition Rules by Graeme Sait *Soil Therapy Pty Ltd 2003*
Chemistry Made Simple by Hess (rev. by Thomas) *Doubleday 1984*
Minerals for the Genetic Code by Charles Walters with Dr. Richard Olree *Acres USA 2006*
Nourishing Traditions by Sally Fallon with Mary Enig *New Trends 2001*
Random House Dictionary of the English Language 2nd Edition Unabridged Flexner and Hauck ed. *Random House 1987*

Highly **Recommended Reading**

Most titles are available from Acres USA www.acresusa.com

Soil Fertility and Animal Health by William A. Abrecht

Eco-Farm: an Acres USA Primer by Charles Walters

Science in Agriculture by Arden Andersen

The Biological Farmer by Gary F. Zimmer

Hands On Agronomy by Neal Kinsey

The Non-Toxic Farming Handbook by Philip Wheeler and Ronald Ward

Nutrition Rules! Guidelines from the Master Consultants by Graeme Sait

Nutrition and Physical Degeneration by Weston A. Price

Nourishing Traditions by Sally Fallon

Bread from Stones by Julius Hensel

Paramagnetism by Phil Callahan

Marschner's Mineral Nutrition of Higher Plants edited by Petra Marschner

Introduction to
Variation in Mineral Composition of Vegetables by Bear, Toth, and Prince
Rutgers University, New Jersey, USA 1948

It is a mark of shame on nutritional and agricultural science that this is the most complete and one of the only comparisons of the mineral values in our food plants as they relate to various soils. Its publication in 1948 marks the high point of health-based agricultural research; from then on the quantity-over-quality industrial model of agriculture took the leading role and soil, plant, animal, and human health have since been largely ignored (see chapter 1 of *The Ideal Soil*).

This is a classic and important study, not only for the data regarding mineral content of vegetables grown in various soils and in different US states, but because it confirms a then century-old dictum of Justus von Liebig, the father of modern soil mineral science. Liebig stated in the 1840s that:

"...the species of one and the same family will contain the same number of basic equivalents combined with vegetable acids."

Von Liebig's principle is restated in this paper as:

"Under uniform conditions for growth, except for limited variations in the relative amounts of the several cations in the nutrient media, the sum of the Ca, Mg K, and Na, expressed in milliequivalents per unit weight of dry matter, is a constant for any given plant variety."

What this means, essentially, is that **each plant species could be said to have its own "cation exchange capacity", and the sum of the various cations that have become part of the plant tissue will have an equal base saturation potential.** (See chapter 2 of *The Ideal Soil*, Cation Exchange Capacity).

Each plant variety will absorb a certain amount of base cation nutrients and no more. If the soil is too high in Potassium, a highly mobile element that plants absorb readily and easily, they may become saturated with K and have no room for Calcium or Magnesium.

The authors also note that **the major anions Sulfur, Chlorine, Phosphorus and Nitrogen appear to have the same sort of sum-total limits of anion saturation**, pointing out the example of a soil with a high level of available nitrate inhibiting phosphate uptake in the crop.

The total amount of cations and anions present in the plant will naturally determine the pH of the plant and its sap. Once the optimum cation/anion concentration and sap pH is attained, which for most plants is pH 6.4, no further cations or anions are readily taken up except for growth of new tissue.

The above examples may have significant impact on the nutritional value of crops. Today's fertilization practices often over-supply Nitrogen and Potassium as both

give a strong growth and yield response, but if one wants to grow and maintain strong bones and teeth, one needs high levels of Calcium and Phosphorus in their food, not high levels of Nitrogen and Potassium.

In the 1939 classic book *Nutrition and Physical Degeneration*, Dr. Weston A. Price showed repeatedly that isolated populations of humans eating their traditional diet consumed about 5 times as much Calcium and Phosphorus as those on a modern diet of industrially grown processed foods, and had none of the modern problems with tooth decay and weak bones.

To those who have studied the writings of William Albrecht on the relationships between climate, soil development, and nutrition it is apparent that a goal of the authors was to provide statistical evidence to bolster Albrecht's observations that the nutrition and health of the population of the USA declines as one goes East from the central states as a consequence of the leaching out of essential mineral elements due to high rainfall over a period of many centuries. This is most apparent in Table 1, which shows that 100% of farmers in the East coast states were using fertilizer, average use being 1500 lbs/acre, while only 39% of the farmers in Colorado used fertilizer, and their average use was only 200 lbs/acre. Table 3 shows that the amount of major nutrient cations was significantly higher across the board in those crops from the East north-central states and Colorado, despite those areas using on average only 1/3 to 1/7 as much fertilizer.

An interesting point about this study is that it has been misused and misinterpreted as a comparison of the mineral nutrition in organic versus chemically grown crops. Even a casual reading will show that it is no such thing, but rather a comparison of the nutritional values of various vegetables grown on different soils in several US states. Attached at the very end of the paper is a 1991 statement from Joseph R. Heckman of the Crop Science Dept. of Rutgers University that makes this point clear.

William A. Albrecht's essay "Our Teeth and Our Soils" is first on the list of citations at the end of the thesis below. What a different world it could have been, but instead here we are sixty years on trying to pick up the pieces of real crop science while at the same time doing damage repair on three-score years of short-sighted greed-based exploitation and poisoning of our agricultural soils. We had best be up to the challenge as there won't likely be another such chance.

Michael Astera
Edited March 11, 2014

Adapted from the original paper at Rutgers University Cooperative Extension Service and the invaluable agricultural reference web site www.soilandhealth.org

Variation in Mineral Composition of Vegetables[1]

FIRMAN E. BEAR, STEPHEN J. TOTH, and ARTHUR L. PRINCE[2]

INTRODUCTION

The percentages of ash and of each constituent in the ash of any given species of plant are known to vary widely. They vary with the variety and with the age of the plant and the environmental conditions under which it was grown. As Sims and Volk have pointed out (9)[3], such variation is of considerable significance to animals and man, since these creatures depend upon plants for most of the mineral matter they require.

Recent studies of plant ash have confirmed Liebig's century-old concept (5) that "the species of one and the same family will contain the same number of basic equivalents combined with vegetable acids." This principle would now be stated as follows:

Under uniform conditions for growth, except for limited variations in the relative amounts of the several cations in the nutrient media, the sum of the Ca, Mg K, and Na, expressed in milliequivalents per unit weight of dry matter, is a constant for any given plant variety.

Recognition that this principle applies in plants was delayed because chemists have long been reporting analyses of plant ash in terms of percentages of the constituent elements, rather than as their equivalents. Within recent years, however, a number of workers have presented their data in equivalent form, and the principle has been adequately confirmed (2, 6, 7). The highest degree of constancy is found in the terminal leaves (10).

Although Ca is the dominant cation in the exchange complex of normal agricultural soils, its rate of movement into the plant is relatively slow in comparison with that of K. Thus, in an experiment with alfalfa (3), it was found that, with a Ca-K equivalent ratio of 32:1 in the exchange complex of the soil, the ratio of these cations in the plants which grew on that soil was only a little over 3:1. This tendency of plants to take up K is such that much larger amounts of it are often absorbed from the soil than are required for optimum crop yields. When this occurs, the absorption of Ca, Mg, and Na is correspondingly reduced. This may be to the disadvantage of the consuming animal and to man.
The principle of constancy also appears to apply to the mineral anions in plants. For example, Nightingale pointed out (8) that application of nitrate results in the reduction of phosphate uptake in pineapples. When soil fumigants were employed and the ammonia forms of nitrogen were not changed to nitrate for a considerable period of time, phosphate absorption was increased.

In a series of alfalfa plants that were grown in our greenhouse under standardized conditions, except for wide variations in the individual anion values in the nutrient media, the sums of the N, S, Cl, and P absorbed, per unit of dry matter, were essentially constant. It should be noted in this connection that the pH values of the nutrient media were kept uniform. This is important in both cation and anion studies that have to do with this point.

Percentages of ash and summation values are known to be subject to wide variations, depending upon the extent to which the dilution factor of carbohydrate production operates. They tend to be considerably higher in the irrigated arid and semi-arid regions than in the more humid regions. This is in conformity with Albrecht's concept of high-carbohydrate versus high protein-and-mineral vegetation regions of the United States (1).

It is apparent from the foregoing that the mineral cation and anion values in plants are an expression of the environment in which the plants were grown. The environmental factors that seem to exert the greatest influence are soil type, fertilizer practice, and climate.

Wide variation in these three environmental factors is readily found as one proceeds from south to north and from east to west in the United States. An opportunity was recently provided[4] to obtain samples of vegetables from a line of states extending northward from Georgia to New York (Long Island) along the Atlantic Coast and from another line of states that extended as far west as Colorado. It is the purpose of this paper to present the results of a study of the mineral composition of the vegetables so selected.
Samples of cabbage, lettuce, snapbeans, spinach, and tomatoes were obtained from commercial fields of these crops in Georgia, South Carolina, Virginia, Maryland, New Jersey, New York (Long Island), Ohio, Indiana, Illinois, and Colorado.[5] The total number of samples examined was 204.

The collecting had to be done during the midsummer months, and this made it impossible to obtain samples of all five crops from all 10 states. Fortunately, samples of snapbeans and tomatoes were taken from every state. This report, therefore, deals primarily with the findings on these two crops. Bountiful snapbeans and Rutgers tomatoes were chosen for collecting and most of the samples belonged to these two varieties. So far as possible, the cabbage, lettuce, and spinach samples were confined to the Golden Acre, Grand Rapids, and Savoy varieties, respectively.

All samples were collected at the stage of growth when they were being harvested for market. Field collection was followed by as rapid transportation to the laboratory as possible. Only the edible portions were prepared for analysis, the outer leaves of cabbage and lettuce being discarded. All samples were rinsed in cold distilled water. The tomatoes were rubbed also with a clean cloth. The samples were dried in a hot-air convection oven at temperatures ranging between 70 and 80° C. Samples of the vegetables were wet-ashed with a mixture of nitric and perchloric acids and made up to volume. Aliquots were then analyzed for the major nutrient elements by standard procedures, including the use of the flame photometer for determining Ca, K, and Na. Another sample was dry-ashed at between 600 and 700° C and analyzed for the minor mineral nutrient elements by the use of a spectrograph.[6]
The soils involved in the eastern coastal-plain states were of the Tifton, Bladen, Orangeburg, Portsmouth, Norfolk, and Sassafras series. These belong to the podzolic group, including both the red-yellow and the gray-brown zones. They have all been developed from coastal-plain materials and have been thoroughly leached, they have relatively low exchange capacities, and they contain only very limited supplies of mineral nutrients.

The soils involved in the east north-central states were of the Wooster, Miami, Crosby, Brookston, Clarion, and Webster series. The first four are members belonging to the gray-brown podzolic group, which have been developed on glacial drift, some of which was of a calcareous nature. Those of the last two series are prairie soils, which have been developed from calcareous glacial drift.

The Colorado vegetables were obtained from areas, where the Laurel, Gilchrist, and Berthan series predominate. These soils belong to the brown and planosol groups, and are under irrigation farming. They are high in calcium carbonate and in available mineral nutrients.
As Beeson has pointed out (4), fertilizing and liming practices influence the mineral composition of plants. Consequently it seemed desirable to make a survey of these practices as employed on the fields from which the samples were selected. The data from this survey are summarized in Table 1. It is important to note the relatively high rates at which fertilizer is applied in the coastal-plain states as compared to the rates employed farther west. In the east north-central states less dependence is placed on fertilizers and greater use is made of clover sods and manure. Only relatively small amounts of fertilizer are used in Colorado.

TABLE 1.
Fertilizer practices in state areas from which vegetable samples were obtained.

	Farmers using fertilizer (%)	**Amount fertilizer per acre (lbs.)**	**Quantities (nutrients per acre)**			**Farmers using sidedressings (%)**
			N (lbs.)	**P_2O_5 (lbs.)**	**K_2O (lbs.)**	
Eastern coastal states	100	1 500	90	120	120	50*
East north-central states	40	500	20	60	50	5**
Colorado	39	200	25	40	10	5**

* Usually nitrate of soda or additional complete fertilizer.
** Some carrier of nitrogen.

Data on the ash and mineral cation content of 46 samples of snapbeans and 67 samples of tomatoes are shown, state by state, in Table 2. Summary values for all five vegetables are given in Table 3.

TABLE 2.
Average ash and nutrient-cation content of snapbeans and tomatoes and highest and lowest individual values for these and three other vegetables. The rate of use of lime increases from Georgia northward to New Jersey. It varies considerably from farm to farm in the east north-central states. No lime was used on the Colorado farms.

Note by m.astera: Apparently the units used to measure cations in the chart below are milliequivalents of H+.

State	Snapbeans					Total cations	Tomatoes					Total cations
	Ash	Ca	Mg	K	Na		Ash	Ca	Mg	K	Na	
Georgia	6.50	14.5	38.3	51.7	1.3	105.8	7.78	6.0	32.9	85.7	3.0	127.6
S.Carolina	6.26	23.0	32.9	44.8	2.2	102.9	8.20	7.5	30.4	85.7	3.5	127.1
Virginia	5.98	17.0	25.5	50.9	1.7	95.1	8.44	7.0	33.7	97.2	2.2	140.1
Maryland	6.49	20.5	36.2	56.0	0.8	113.5	7.00	14.0	14.8	88.2	0.4	117.4
NewJersey	6.62	24.0	43.6	48.8	3.9	120.3	8.14	13.0	21.4	83.1	2.2	119.7
New York	6.34	25.5	39.5	64.5	3.0	132.5	8.95	14.5	17.3	107.4	1.3	140.5
Ohio	8.53	30.5	45.2	71.1	1.7	149.0	9.10	13.5	26.3	101.8	1.3	142.6
Indiana	6.59	30.5	46.0	67.5	1.3	145.3	9.18	15.0	28.0	101.8	2.2	147.0
Illinois	7.73	26.5	43.6	70.6	1.3	142.0	8.59	13.8	28.0	96.0	1.3	139.1
Colorado	7.68	29.0	48.5	56.5	0.4	134.4	11.54	15.0	33.7	111.0	0.8	160.5
	Snapbeans						**Tomatoes**					
Highest	10.45	40.5	60.0	99.7	8.6		14.20	23.0	59.2	148.3	6.5	
Lowest	4.04	15.5	14.8	29.1	0.0		6.07	4.5	4.5	58.8	0.0	
	Cabbage						**Spinach**					
Highest	10.38	66.0	43.6	148.3	20.4		28.56	96.0	203.9	257.0	69.5	
Lowest	6.12	17.5	15.6	53.7	0.8		12.38	47.5	46.9	84.6	0.8	
	Lettuce											
Highest	24.28	71.0	49.3	176.5	12.2							
Lowest	7.01	16.0	13.1	53.7	0.0							

*No two of these extreme values are for the same sample. Thus, for snapbeans the highest Ca, Mg, K, and Na values were found in Colorado, Colorado, Indiana, and New York, respectively.

TABLE 3.
Average percentages ash, and macronutrients in dry matter*
of five vegetables grown on eastern coastal plain soils and on east north-central states and Colorado soils.

	Snapbeans		**Tomatoes**		**Cabbage**		**Lettuce**		**Spinach**	
	Eastern coastal states	**East north central states and Colorado**	**Eastern coastal states**	**East north central states and Colorado**	**Eastern coastal states**	**East north central states and Colorado**	**Eastern coastal states**	**East north central states and Colorado**	**Eastern coastal states**	**East north central states and Colorado**
Ash	6.38	7.63	8.08	9.59	8.64	7.77	9.63	13.34	23.63	28.61
Ca	0.43	0.58	0.20	0.27	0.59	0.70	0.46	0.87	1.30	1.57
Mg	0.45	0.55	0.30	0.35	0.35	0.32	0.35	0.43	1.30	1.83
K	2.12	2.59	3.56	3.96	2.71	2.64	3.55	4.85	7.38	7.34
Na	0.05	0.02	0.05	0.03	0.20	0.07	0.17	0.04	0.46	0.78
P	0.25	0.25	0.24	0.30	0.28	0.22	0.34	0.32	0.37	0.30

*The percentages dry matter in the fresh vegetables averaged 8, 6, 7, 5, and 10 for snapbeans, tomatoes, cabbage, lettuce, and spinach, respectively.

After consideration of the state-average and summary values, in conjunction with the individual values for the 204 samples of all five vegetables, of which only the extremes are shown at the bottom of the table, the following conclusions were drawn:

1. Ash, Ca, and cation-equivalent values tend to increase from south to north and from east to west.
2. K values tend to increase from east to west.
3. Mg values tend to increase from north to south and from east to west.
4. Na values tend to decrease from east to west.[7]

TABLE 4.
Average phosphorus and minor nutrient content of snapbeans and tomatoes and highest and lowest individual values* for these and three other vegetables. P in percentage and minor elements in parts per million dry matter.

State	Snapbeans							Tomatoes						
	P	**B**	**Mn**	**Fe**	**Mo**	**Cu**	**Co**	**P**	**B**	**Mn**	**Fe**	**Mo**	**Cu**	**Co**
Georgia	0.27	14	24	83	0.5	12	0.02	0.25	8	6	107	0.1	10	0.03
S. Carolina	0.27	17	9	110	0.4	13	0.05	0.27	10	4	119	0.1	11	0.06
Virginia	0.28	12	21	68	0.1	17	0.05	0.27	7	3	59	0.2	21	0.01
Maryland	0.22	12	30	75	0.2	11	0.12	0.19	10	5	97	0.1	16	0.04
New Jersey	0.25	25	7	88	0.6	14	0.03	0.24	9	7	113	0.2	20	0.08
New York	0.23	16	20	74	0.5	9	0.06	0.23	11	2	87	0.1	26	0.4
Ohio	0.27	15	14	77	3.0	16	0.06	0.27	30	3	96	0.3	12	0.02
Indiana	0.24	20	7	130	5.0	14	0.03	0.29	12	4	52	0.5	14	0.06
Illinois	0.25	19	7	129	3.4	30	0.05	0.30	12	2	179	2.0	27	0.03
Colorado	0.26	16	4	130	4.3	24	0.06	0.25	13	4	265	0.5	24	0.11
	Snapbeans							**Tomatoes**						
Highest	0.36	73	60	227	8.1	69	0.26	0.35	36	68	1,938	1.3	53	0.63
Lowest	0.22	10	2	10	0.1	3	0.00	0.16	5	1	1	0.0	0	0.00
	Cabbage							**Spinach**						
Highest	0.38	42	13	94	24.1	48	0.15	0.52	88	117	1,584	5.6	32	0.25
Lowest	0.18	7	2	20	0.0	0.4	0.00	0.27	12	1	19	0.0	0.5	0.20
	Lettuce													
Highest	0.43	37	169	516	4.5	60	0.19							
Lowest	0.22	6	1	9	0.0	3	0.00							

*See note at bottom of Table 2.

The P, B, Mn, Fe, Mo, Cu, and Co content of the same samples of snapbeans and tomatoes from all 10 states are shown in Table 4. Studies of these state average values, in conjunction with the 204 individual values, of which only the extremes are shown at the bottom of the table, permit of the following conclusions:

1. P values are relatively constant from state to state, but the individual values for each vegetable vary between wide extremes.
2. B, Fe, Mo, Cu, and Co values tend to increase from east to west.
3. Mn values tend to decrease from east to west.

Wide variations were found from region to region in the percentage ash and of each of the individual mineral nutrient elements in the ash.

Wide variations were found in the cation-summation values. This is to be expected, since the environmental conditions under which the plants had been grown were very dissimilar.

Spinach was notably high in ash. Variations in K, Na, B, and Fe values were greatest in this plant. The K values varied between 10.05 and 3.31%, the Na values between 1.60 and 0.02%, the B values between 88 and 12 ppm, and the Fe values between 1584 and 19 ppm.[8] Spinach appeared to be an accumulator of both Mo and Co.

Tomatoes showed the greatest variation in Ca, Mg, and Cu. The Ca values varied between 0.40 and 0.09%, the Mg values between 0.72 and 0.14%, and the Cu values between 46 and 0 ppm.

Snapbeans grown in Ohio, Indiana, Illinois, and Colorado were notably high in Mo. The average Mo value for the four east north-central states and Colorado was 3.9 ppm, in comparison with 0.4 ppm for the six coastal-plain states. The highest Mo value, 24.1 ppm, was found in a sample of Indiana cabbage.

Lettuce and spinach were two exceptions in the general trend of higher Mn values in the eastern states than in the east north-central states and Colorado. The explanation for this probably lies in the fact that eastern soils are usually well limed for these crops. Often they are overlimed. The lowest Mn value, 0.6 ppm, was found in a sample of lettuce from New Jersey, and the highest, 161 ppm, in a sample from Indiana.

Colorado vegetables, in comparison with those from the other nine states, were relatively high in Co, Mo, Cu, and Ca in the order indicated. They were moderately high in K, Mg, Fe, and B, in the order indicated. They were about average in P, relatively low in Mn, and very low in Na.

The K content of Colorado vegetables was not as high relatively as one might expect. The explanation for this is found in the fact that the soils of Colorado are relatively very high in Ca and Mg, as well as in K. It is important to note also that liberal applications of K, in the form of fertilizers and manures, are made to the land in the east and south in preparation for growing vegetables. This is in marked contrast to the very small rates of application of such materials in Colorado.

SUMMARY AND CONCLUSIONS

Two hundred and four samples of cabbage, lettuce, snapbeans, spinach and tomatoes were analyzed for their content of ash, Ca, Mg, K, Na, P, B, Mn, Fe, Mo, Cu, and Co.

These samples were chosen from Georgia, Virginia, South Carolina, Maryland, New Jersey, New York (Long Island), Ohio, Indiana, Illinois, and Colorado.

Wide variations were found in the mineral content of vegetables of the same variety.

Ash, Ca, and cation-equivalent values tended to increase and Mg values to decrease from south to north.

Ash, cation-equivalent, Ca, Mg, K, B, Fe, Mo, Cu, and Co values tended to increase from east to west.

Na and Mn values tended to decrease from east to west.

P values tended to be relatively constant, but wide individual variations were found in the same variety of vegetable.

The greatest variations in K, Na, B, and Fe values were found in spinach.

The greatest variations in Ca, Mg, and Cu values were found in tomatoes.

Snapbeans from Ohio westward were relatively very high in Mo.

Colorado vegetables, in comparison with those from the other states, were relatively high in Co, Mo, Fe, Ca, K, Mg, Cu, and B, in the order indicated; about average in P; and relatively low in Mn and Na.

LITERATURE CITED

1. ALBRECHT, Wm. A. Our teeth and our soils. Ann. Dent., 6:199-213. 1947.
2. BEAR, FIRMAN E., and PRINCE, ARTHUR L. Cation-equivalent constancy in alfalfa. Jour. Amer. Soc. Agron., 37:217-222. 1945.
3. ------ and TOTH, STEPHEN J. Influence of calcium on availability of other soil cations. Soil Sci., 65:69-74. 1948.
4. BEESON, KENNETH C. The mineral composition of crops, with particular reference to the soil on which they w'ere grown. U.S.D.A. Misc. Pub, No. 369. 1941.
5. BOUSSINGAULT, J. B. Rural Economy. Translation bv George Law. New York: Orange Judd. p. 64. 1865.
6. HARMER, PAUL M., and BENNE, ERWIN J. Sodium as a crop nutrient. Soil Sci., 60: 137-149. 1945.
7. LUCAS, R. E., and SCARSETH, G. D. Potassium, calcium. and magnesium balance and reciprocal relationships in plants. Jour. Amer. Soc. Agron., 39: 887-897. 1947.
8. NIGHTINGALE, GORDON T. The nitrogen nutrition of green plants. 11. Bot. Rev., 14: 185-221. 1948.
9. Sims, G. T., and VOLK, G. M. Composition of Florida-grown vegetables. Fla. Agr. Exp. Sta. Bul. 438. 1947.
10. WALLACE, ARTHUR, TOTH, STEPHEN J., and BEAR, FIRMAN E. Further evidence supporting cation-equivalent constancy in alfalfa. Jour. Amer. Soc. Agron., 40: 80-88. 1948.

Misquotes in "Variation in Mineral Composition of Vegetables"

A study conducted at Rutgers University (Bear et al., 1948) is frequently misquoted as evidence supporting the position that organically grown vegetables are significantly superior in minerals and trace elements to conventionally grown vegetables. In reviewing the original publication, one can clearly see that this was not the intention of the study nor does it give support to this premise. The purpose of the study was to compare the mineral composition of vegetables "as one proceeds from south to north and from east to west in the United States." Samples of cabbage, lettuce, snapbean, spinach, and tomatoe were obtained from commercial fields of these crops and analyzed for mineral composition. A total of 204 samples were examined. The vegetables sampled were usually, but not always, of the same variety. The authors reported, in a table, the range in mineral concentration as highest and lowest values observed among the vegetables sampled. These highest and lowest values have been misrepresented as vegetables grown organically and inorganically, respectively, in various organic farming and healthfood newsletters, which cite the report (copies of the misquotes are available on request).

The authors discussed the influence of soil type, fertilizer practice, and climate on the observed differences in mineral composition. The study only provides a general survey of their possible influence and did not compare synthetic fertilizer and organic practices.

Received 11 Mar. 1991.

JOSEPH R. HECKMAN
Crop Science Dept.
Rutgers Univ.
New Brunswick, NJ 08903

References

Bear, F.E., S.J. Toth, and A.L. Prince. 1948. Variation in mineral composition of vegetables. Soil Sci. Soc. Am. Proc. 13:380-384.

Reprinted from the *Soil Science Society of America Journal*
Volume 55, No. 5, September-October 1991
677 South Segoe Rd., Madison, WI 53711 USA

Primitive Diets: Weston A. Price's findings on the daily intake of the major nutrients Calcium, Magnesium, Phosphorus, and Iron among isolated groups consuming traditional foods, 1939

Following are some excerpts from *Nutrition and Physical Degeneration*, along with a table compiled from data in the same book, showing the amounts of major nutrients that Price measured in the typical diets of these isolated groups.
Notes in brackets [ ] are by m.astera.

~~~~~~~~~~~~~~~~~~~

In WA Price's *Nutrition and Physical Degeneration*, pp 274-276 he writes:

"It will therefore be necessary for an adequate nutrition to contain approximately four times the minimum requirements of the average adult if all stress periods are to be passed safely."

"If we use as a basis the ability of individuals to remove half of the minerals present even though their bodies need more than this, we will be more generous than the average individual's capacity will justify. This will require that we double the amount, as specified for minimum body use by the United States Department of Labor, bureau of Labor Statistics, in their Bulletin R 409, that is, for calcium 0.68 grams; for phosphorus 1.32 grams; for iron 0.015 grams. The figures that will be used, therefore, are for twice the above amounts: 1.36 grams of calcium, 2.64 grams of phosphorus; 0.030 grams of iron.

It is of interest that the diets of the primitive groups which have shown a very high immunity to dental caries and freedom from other degenerative processes have all provided a nutrition containing at least four times these minimum requirements."

****

Price indicates above that the figures he is using are 2x the 1930s US Department of Labor recommendations, which were:
Ca = 1360mg
P = 2640mg
Fe = 30mg

2008 USDA RDAs (Recommended Daily Alllowances) are
Ca = 1300mg
P = 1250mg
Fe = 27mg
(note that these are the *maximum* RDAs from all categories of the USDA chart)

If we use Price's minimums for Calcium of 1360mg x 5.1 we get 6936mg Ca.
For Phosphorus, 2640mg x 5.4 = 14 256mg.

Who in today's world is getting 7grams of Calcium and over 14 grams of Phosphorus per day? How does one manage that on a 2000 calorie diet?

I would suggest that this P level could only be achieved with the sort of soil phosphate (P_2O_5) levels that Carey Reams recommends, i.e. 2x potash (K_2O) for most crops and 4x potash for grasses and legumes.
~~~~~~~~~~~~~~~~~~~

Comparison of Mineral Nutrients in Traditional Diets as a Multiple of 2x US Labor Dept. Recommendations [1930s]

[As I interpret Price's words above, he ***doubled*** the minimum recommendations of the USDL, then calculated how many times more nutrients these isolated populations were consuming, i.e. where it says the Eskimos were getting 5x as much Phosphorus, they were actually getting 10x the USDL minimum recommendations.]

Group	Calcium	Magnesium	Phosphorus	Iron	Copper	Iodine	Fat soluble vitamins
Eskimos	5.4x	7.9x	5x	1.5x	1.8x	49x	>10x
N. Canadian Indians	5.8	4.3	5.8	2.7	1.5	8.8	>10
Swiss Alps	3.7	2.5	2.2	3.1			>10
Outer Hebrides Gaelics	2.1	1.3	2.3	1.0			>10
E. Coast Australian Aborigines	4.6	17	6.2	50.6			>10
NZ Maori	6.2	23.4	6.9	58.3			>10
Melanesians	5.7	26.4	6.4	22.4			>10
Polynesians	5.6	28.5	7.2	18.6			>10
Peru Coastal Indians	6.6	13.6	5.4	5.1			>10
Peru Andes Indians	5	13.3	5.5	29.3			>10
Central African Cattle Tibes	7.5	19.1	8.2	16.6			>10
Central African Agricultural Tribes	3.4	5.4	4.1	16.6			10
Average	5.1x	13.6x	5.4x	17.4x			>10x
	Calcium	Magnesium	Phosphorus	Iron			Fat Soluble Vitamins

"All of the above primitive diets provided also a large increase in the water-soluble vitamins over the number provided in the displacing modern diets."
Table compiled from data in W.A. Price *Nutrition and Physical Degeneration* 1939 pp 274-276

Calculating Total Cation Exchange Capacity TCEC

The simplest formula for calculating CEC is

$$\frac{\text{ppm Ca}}{200} + \frac{\text{ppm Mg}}{120} + \frac{\text{ppm K}}{390} + \frac{\text{ppm Na}}{230} = \text{CEC in meq}$$

The formula above can be used as-is for calculating the CEC of **soils of pH >7** but only when using the results of an Ammonium Acetate pH8.2 test. See Chap. 9.

Most soils with a pH<7 will have other base cations besides Ca, Mg, K, and Na occupying negative exchange sites, for instance ammonium NH_4+, Fe+. Cu+, along with acidic ions such as H+ and Al^{3+}. If we don't know how much of the exchange capacity is occupied by other bases and H+, we will not be able to estimate the true CEC of the soil.

At some time during the 1960s, '70s, or '80s, unknown researchers, probably working at Brookside Laboratories in Ohio, developed a modification of the formula above that allowed them to make a close estimation of these other bases and acidic cations, using correction factors based on the pH of the soil. CEC calculated in this way is called Total Cation Exchange Capacity, TCEC, because it includes an estimate of the amounts of other cations held on exchange sites.

Experience has shown that the Brookside formula for calculating TCEC is very accurate when applied to soils having a pH<7 using the results of a Mehlich 3 test.

The Brookside Labs formula for calculating Total Cation Exchange Capacity:

$$\frac{\frac{\text{ppm Ca*}}{200} + \frac{\text{ppm Mg*}}{120} + \frac{\text{ppm K*}}{390} + \frac{\text{ppm Na*}}{230}}{100 - (\text{other bases} + \text{exchangeable H+})} \times 100 = \text{TCEC}$$

Other Bases:

= 0	if	soil pH > 7.0
= 11.4 - pH	if	soil pH > 6.1 ≤ 7.0
= 17.4 - (2 x pH)	if	soil pH > 3.0 & ≤ 6.1
= 13.3 - (.6 x pH)	if	soil pH ≥ 2.2 & ≤ 3
= 17.4 - (2 x pH)	if	soil pH < 2.2

Exchangeable Hydrogen:

= 0	if	pH > 7.0
= (7 - pH) x 15	if	pH > 6.0 & ≤ 7.0
= 195 - (30 x pH)	if	pH > 5 & ≤ 6.0
= 145 - (20 x pH)	if	pH > 4.0 & ≤ 5.0
= 105 - (10 x pH)	if	pH > 3.0 & ≤ 4.0

= 93 - (6 x pH) if pH ≥ 2.2 & ≤ 3.0
= 155 - (25 x pH) if pH < 2.2

The Brookside TCEC formula may look a bit complex at first, but we will break it down into simple pieces.

$$\frac{\frac{\text{ppm Ca*}}{200} + \frac{\text{ppm Mg*}}{120} + \frac{\text{ppm K*}}{390} + \frac{\text{ppm Na*}}{230}}{100 - (\text{other bases} + \text{exchangeable H+})} \times 100 = \text{TCEC}$$

If you have read chapter 2 on Cation Exchange the elements and numbers above should be familiar.

In the top part of the equation, the lab results in ppm for the major cations Ca, Mg, K, and Na are each being divided by the amount in ppm that would be needed to saturate 100% of the exchange sites on a soil with a CEC of 1.
1 meq Ca= 200ppm, 1 meq Mg = 120ppm etc.

If Ca, Mg, K, and Na were the only cations present, the CEC of the soil would simply be the sum of those results. For example, imagine our lab results show 200ppm Ca, 120ppm Mg, 390ppm K, and 230ppm Na.

Ca 200ppm / 200 = 1meq
Mg 120ppm / 120 = 1meq
K 390ppm / 390 = 1meq
Na 230ppm / 230 = 1meq

The CEC will be 1 + 1 + 1 + 1 = 4meq, and the base saturation % of each cation will be the amount reported by the lab test divided by the amount needed to saturate 100% of the exchange sites in a soil with a CEC of 4meq:

Ca 200ppm / (4 x 200) = 0.25 or 25%
Mg 120ppm / (4 x 120) = 0.25 or 25%
K 390ppm / (4 x 390) = 0.25 or 25%
Na 230ppm / (4 x 230) = 0.25 or 25%

Our imaginary test numbers were exactly the amount needed to saturate 1meq of cation exchange capacity for each base cation. Added together they indicate a CEC of 4meq, with each one of these base cations filling 25% of the CEC.

Next, the same calculations done using the numbers from the soil test that we have been working with throughout the book:

Element		Results
Cation Exchange Capacity CEC meq		11.4
pH of Soil Sample		5.58
Primary Cations		
Calcium Ca++ ppm	Desired Found Deficit	1550 1250 -300
Ca Base Saturation 60-70 %		55.0%
Magnesium Mg++ ppm	Desired Found Deficit	164 116 -48
Mg Base Saturation 10-20 %		8.5%
Potassium K+ ppm	Desired Found Deficit	178 89 -89
K Base Saturation 2-5 %		2.0%
Sodium Na+ ppm	Desired Found Deficit	39 26 -13
Na Base Saturation 1-5 %		1.0%

Using the "Found" numbers from this table for the first part of our TCEC calculation:

$\frac{\text{Ca 1250ppm}}{200} = 6.250$

$\frac{\text{Mg 116ppm}}{120} = 0.967$

$\frac{\text{K 89ppm}}{390} = 0.228$

$\frac{\text{Na 26ppm}}{230} = 0.113$

6.25 + 0.97 + 0.23 + 0.11 = 7.56 CEC. This number is what many laboratories would report for the CEC of the soil sample. (Some labs and agronomy texts only include Ca, Mg, and K in the calculations, not Na.)

The soil report we have been working with says our exchange capacity is 11.4meq. Why the difference? Because there are other cations filling exchange sites in this soil. We need to calculate how much of the Total CEC is taken up by these other cations.

That's where the divisor in the TCEC equation comes into play; we need to take the 7.56 number we found above and divide it by:

100 - (other bases + exchangeable H+)

Here is how the "other bases" are calculated:

% Other Bases:

= 0		if	soil pH >7.0
= 11.4	- pH	if	soil pH > 6.0f & $\leq$ 7.0
= 17.4	- (2 x pH)	if	soil pH > 3.0 & $\leq$ 6.0
= 13.3	- (0.6 x pH)	if	soil pH $\geq$ 2.2 & $\leq$ 3
= 17.4	- (2 x pH)	if	soil pH < 2.2

Our pH is 5.58, which is > (greater than) 3.0 and ≤ (less than or equal to) 6.1. We use the numbers from the second row above:

17.4 - (2 x 5.58pH) or 17.4 - 11.16 = 6.24

So we know that “other bases” equal 6.24% of CEC. Next we need to calculate exchangeable Hydrogen H+:

% Exchangeable Hydrogen:

= 0	if	pH > 7.0
= (7 - pH) x 15	if	pH > 6.0 & ≤ 7.0
= 195 - (30 x pH)	if	pH > 5.0 & ≤ 6.0
= 145 - (20 x pH)	if	pH > 4.0 & ≤ 5.0
= 105 - (10 x pH)	if	pH > 3.0 & ≤ 4.0
= 93 - (6 x pH)	if	pH ≥ 2.2 & ≤ 3.0
= 155 - (25 x pH)	if	pH < 2.2

Our pH, 5.58, is > 5 and ≤ 6.0. We use the figures from row 3 above:

195 - (30 x 5.58) or 195 - 167.4 = 27.6

At pH 5.58, the CEC will be saturated 27.6% with H+ and 6.24% with other bases. Filling in the blanks we get:

100% - (6.24% other bases + 27.6% exchangeable H+)

100% - 33.84% = 66.16 or 66.2%

66.2% of the CEC of this soil is saturated with Ca, Mg, K, and Na ions, 33.8% with other bases and exchangeable Hydrogen.

Back to the first number we calculated from the top line of the formula, 7.56. We want to know what number 7.56 is 66.2% of:

7.56 / 66.2 = 0.114 Next, we take that number x 100

0.114 x 100 = 11.4 TCEC.

11.4 is the Total Cation Exchange Capacity; it includes any other cations held on the exchange sites. That number is what everything else on the soil report and on The Ideal Soil chart is calculated from and in reference to. If we had just used the 7.56 figure we came up with first, which many labs and agronomists do, assuming that is 100% of CEC, our base saturation numbers would look like this:

Ca: 1250ppm / (7.56 x 200) = 82.7%
Mg: 116ppm / (7.56 x 120) = 12.8%
K: 89ppm / (7.56 x 390) = 3.0%
Na: 26ppm / (7.56 x 230) = 1.5% Note that these add to100%

Using the correctly adjusted TCEC, 11.4meq, our percentages are:

Ca: 1250ppm / (11.4 x 200) = 54.8%
Mg: 116ppm / (11.4 x 120) = 8.5%
K: 89ppm / (11.4 x 390) = 2.0%
Na: 26ppm / (11.4 x 230) = 1.0% Note that these add to 66.3% of base saturation.

Instead of CEC of 7.56meq, this soil has a TCEC of 11.4meq, 50% higher. Using the first numbers, we would think this soil was high in Ca at ~ 83% saturation and had an almost ideal amount of Mg at about 13%, instead of the true Ca saturation of 55.3% and Mg saturation of 8.6%.

If your favorite soil testing lab is not using this method to calculate CEC you might consider asking them to start using it on the tests they do for you. At the very least it would be a good idea to check the lab's CEC and Ca% numbers against the CEC you arrive at using the TCEC formula.

Estimating Nitrogen Release ENR from Soil Organic Matter and Protein

Most plant-available Nitrogen in natural systems comes from the breakdown of soil organic matter. Soil Organic Matter (SOM or OM) may be defined as "the organic fraction of the soil exclusive of undecayed plant and animal residues". Which means freshly fallen and undecayed leaves, straw, animal remains, and living things don't count as part of the soil organic matter. Anything once living but now in the process of decomposing can be considered part of the SOM.

The soil report we have been working with in The Ideal Soil Handbook shows 5.6% OM. We assume that the top 6 to 7 inches (15-17 cm) of soil weighs 2 000 000 lbs/Acre or 2 000 000 kg/hectare, written as 2 000 000# in this book.

2 000 000# x 0.056 = 112 000# organic matter.

The usual assumption is that **SOM contains 5% Nitrogen by weight**, so how much N is potentially available in this soil?

112 000# x 0.05 = 5600# Nitrogen

Most crops need around 80 to 100# of N during a growing season; some such as corn (maize) will use a lot more if it is available, up to 300#. It would seem we have plenty of potential N in this soil, but of course not all of that organic matter is going to break down and become available quickly. Soil temperature, pH, mineral balance, available moisture, and biological activity as well as the addition of other N sources are the controlling factors for how rapidly the OM breaks down and how much N is released during the crop's growing season.

US Dept. of Agriculture figures estimate that 15 to 25# (average 20#), of N are released for growth purposes per year for each 1% OM in temperate climates. Our 5.6% OM content should give us

20# x 5.6% OM = 112# of N

That should be plenty for most crops, if the other factors like temperature and moisture cooperate. In soils with a pH below 6 or above 8, or heavy dense soils that have a high amount of Magnesium and little Oxygen, we would expect less. Soils with a perfect cation balance, sufficient nutrient anions like Sulfur and Phosphorus, abundant trace minerals, and strong biological activity could probably supply double that amount or more, and do it without lessening the OM or N reserves because the soil microbes would be "fixing" as much or more N from the atmosphere as the plants were taking up.

For those interested in some more precise calculations of ENR, here are some equations that correlate the soil OM % with N release. Their original source is unknown, but our best guess is that like the TCEC formulas discussed elsewhere they originated from William Albrecht's collaboration with Brookside Laboratories

---

Estimated Nitrogen Release "units #" (lbs/acre or kg/ha)

= 20 + [(OM % – 0.5) x 40]	if OM is < 1%
= 40 + [(OM % – 1) x 20]	if OM > 1% and < 3%
= 80 + [(OM % – 3) x 10]	if OM > 3% and < 5%
= 100 + [(OM %– 5) x 5]	if OM > 5% and < 10%
= 125 + [(OM %– 10) x 0.5]	if OM > 10% and < 20%
= >130 lbs of N	if OM% is > 20%

Let's try this formula with our 5.6% SOM number from the soil report, using the fourth line above because 5.6% is > (greater than) 5% and < (less than) 10%:

100 + [(5.6 – 5) x 5] or

100 + (0.6 x 5) = 103# ENR; less than our previous rough estimate of 112#.

What happens if we calculate ENR using the "ideal" 4% SOM, >3% and <5%?

80 + [(4 – 3) x 10] or

80 + (1 x 10) = 90# ENR

Estimating Nitrogen Release from Seed Meals and Animal Byproducts:
The standard method of estimating the Nitrogen content of protein is called the Kjeldahl method. This empirical factor is based on a Nitrogen content in protein of 16g N per 100g of protein.

100 / 16 = 6.25

The conversion is applied both ways. If one knows the N content, that N number is multiplied by 6.25 to get the estimated protein content. If the protein content is known, the protein number is divided by 6.25 or multiplied by 0.16 to get the estimated Nitrogen content. For example:
15% protein / 6.25 = 2.4% N

15% protein x 0.16 = 2.4% N

2.4% N x 6.25 = 15% protein

Dealing with Excess: Alkaline and High-Sodium Soils, and When There is Already Too Much

We have the same tools to work with whether the goal is to balance an out of balance acidic soil, to lower the pH of alkaline soils or to reclaim soils that are too high in Sodium or other salts. Our tools are the elements Calcium, Magnesium, Potassium, and Sulfur, and the “universal solvent” H_2O, water.

The goal is always to get the soil’s exchange sites to let go of the element we don’t want and replace it with the element we do want.

The question becomes, which elements will replace which other elements? There are three factors involved:

1. Soil exchange sites have a higher affinity for **divalent** (double ++ charged) cation elements like Ca++ and Mg++ than they do for **monovalent** (single + charged) elements like K+ and Na+.

Mg and Ca will have a stronger attraction to negative exchange sites and tend to replace Na and K on - exchange sites because they have twice the + charge as Na and K do. The second factor involved is:

2. An element with a smaller **hydration radius** will replace one with a larger **hydration radius**.

When a + cation is free in the soil it will attract free H_2O water molecules; that is called hydration. The smaller the atomic size of the ion the larger the diameter of the cluster of water molecules it will attract . Why? Because the water molecules (which have a slight negative polarity on one side) will cluster in a thick layer around the smaller ion, but in a thinner layer around a larger ion of the same charge.

The Calcium++ ion has an atomic number of 40 and is larger than the Magnesium ion, atomic number 24. Because they both have the same ++ charge they may attract the same number of water molecules, but those water molecules will form a smaller diameter cluster around the larger Ca++ ion than around the smaller Mg++ ion.

For the same reason, the size of the ion, the smaller Sodium+ ion (at wt 23) will form a larger (thicker) cluster of water molecules than the larger Potassium+ ion (at wt 39). They both have the same charge and both will tend to attract the same amount of H_2O molecules, but those H_2O molecules will form a “thicker” cluster around the smaller Na+ ion than around the larger K+ ion.

Here are the sizes (radii) of these cations in Angstrom units(□). (An Angstrom unit is 10^{-10} meters or 0.0000000001 meters, or 0.00000000357 inches.

Element	Non-hydrated	Hydrated
Potassium K+	2.66	7.6
Sodium Na+	1.90	11.2
Calcium ++	1.98	19.2
Magnesium ++	1.30	21.6

The size of the hydrated ions, in increasing order, is K+ < Na+ < Ca++ < Mg++

Looking at size alone, we would guess that because it is has the smallest radius of hydration, the K+ ion would tend to displace the other ions because it could get closer to the exchange site. This is true for ions with the same charge. K+ will replace Na+ because K+ has a smaller radius of hydration. Ca++ will replace Mg++ because it has a smaller radius of hydration.

However, as was stated above, a divalent ++ ion has a stronger attraction to an exchange site than a monovalent + ion, so what will happen in practice is that Ca++ will have a stronger attraction to an exchange site than the other three ions listed.

Here is the order of attraction/replacement of these four major cation elements:

Ca > Mg > K > Na

Calcium++ will replace Magnesium++ which will replace Potassium+ which will replace Sodium+.

This holds true when there are equal concentrations of the ions. The third factor affecting ion adsorbtion is

3. **The higher the concentration** of a given cation in the soil/water solution, **the more it will tend to displace other cations** from exchange sites.

If we pour a solution containing 50ppm each of Ca, Mg, K, and Na onto soil with 50 empty exchange sites, the exchange sites will first attract and hold a Calcium ion, but as there become fewer Calcium ions in solution, the Magnesium ion concentration increases and Mg will start to fill some sites. As the Mg is pulled out of solution, K will be next to be attracted and held.

If we want to replace K with Mg, for instance, we would want to provide a greater concentration of Mg in the soil solution.

Alkaline Soils (above pH 7)

In an acid soil, one with a pH below 7, some of the exchange sites will be occupied by the "acidic" cation Hydrogen H+. Recall that in the term pH the H stands for Hydrogen. pH is the measure of the ratio of Hydrogen H+ ions to hydroxide OH- ions. A water solution that has more H+ than OH- ions is acid; if there are more OH- ions than H+ ions, it is alkaline. At pH 7, H+ and OH- are equally balanced.

We are talking about replacing or exchanging cations, + charged ions, on the soil colloid. A soil with a pH of 6 will have around 15% of the cation exchange sites occupied by H+ ions. In a soil with a pH of 7, all of the negative exchange sites will be filled by cations other than H+. The H+ concentration on the exchange sites at pH 7 and above is 0.00%.

Balancing the nutrient cations in an acid soil is relatively straightforward: simply replace some of the H+ ions with the nutrient cations that we wish to see. If the soil has a 60% Calcium saturation and we want a 70% Calcium saturation, we add enough Ca++ to raise the saturation 10%. This may need to be done a few times as the process is not 100% efficient but it will work.

A soil with a pH of 8 is a different story. There are no H+ ions to replace; adding more Ca++ will only make the soil more alkaline. What to do? We need to use an acidifier along with the cation base that we wish to replace. That acidifier is Sulfur, either in the form of elemental Sulfur S, or combined with one of our cation nutrients as a sulfate: Calcium sulfate $CaSO_4$, Magnesium sulfate $MgSO_4$, or Potassium sulfate K_2SO_4. Which one we use will depend on the present level of Sulfur in the soil and of course on which element we wish to raise or lower.

Recall from the Ideal Soil Chart that for most crops we want Sulfur S to be around 50% of Ideal Potassium K. If Ideal K is 200 ppm, Sulfur should be around 100 ppm. If our soil test says we have only 50 ppm S, then we know that we can add 50 ppm S and still be within the Ideal range.

In any soil, we simply look at the level of S and the level of the Nutrient cation that we wish to add or change. If the soil is low in Mg and S, we would add Magnesium sulfate; if it is low in K and S we would use Potassium sulfate. If it has adequate or excessive S we would not use the sulfate form, we would use the oxide or carbonate form, e.g. Calcium carbonate or Magnesium oxide.

In an alkaline soil that had too much Magnesium or Potassium or Sodium, our choices would be limited to using either elemental Sulfur S or Calcium sulfate $CaSO_4$, commonly called gypsum. If an alkaline soil already had plenty of Calcium, we would choose pure elemental Sulfur, or 90%S agricultural sulfur.

Alkaline soils can do fine with much higher levels of Calcium than are considered ideal for a more acid soil. 80 to 85% Calcium, 8 to 10% Magnesium, and 4 to 6% Potassium is a good cation balance for soils with pH > 7.2. One way of looking at this would be to say that the percent of Hydrogen+ that would be found in an acid soil would be replaced with the same percentage of Calcium in an alkaline soil.

If there is an excess of Sodium Na, our goal will be to bring the level down by replacing Na. In most cases the Na would be exchanged for Ca, and the primary tool would be gypsum, Calcium sulfate.

This next part, about the uses and properties of gypsum, is adapted from the website of the USA Gypsum company. We think it is very well done, and trust they won't mind our reprinting it and promoting their products.

Agricultural Gypsum Uses (from: http://www.usagypsum.com/agricultural-gypsum.aspx)

Agricultural Gypsum (Calcium Sulfate - $CaSO_4$) is one of those rare materials that performs in all three categories of soil treatment: an amendment, a conditioner, and a fertilizer.

Poor soil structure is a major limiting factor in crop yield.

Gypsum Improves Compacted Soil
Gypsum can help break up compacted soil. Soil compaction can be prevented by not plowing or driving machinery on soil when it is too wet. The compaction in many but not all soils can be decreased with gypsum, especially when combined with deep tillage to break up the compaction. Combining gypsum with organic amendments also helps, especially in preventing return of the compactions.

Gypsum Helps Reclaim Sodic Soils
Gypsum is used in the reclamation of sodic soils. Where the exchangeable Sodium percentage (ESP) of sodic soils is too high, it must be decreased for soil improvement and better crop growth. The most economical way is to add gypsum which supplies Calcium. The Calcium replaces the Sodium held on the clay-binding sites. The Sodium can then be leached from the soil as Sodium sulfate to an appropriate sink. The sulfate is the residue from the gypsum. Without gypsum, the soil would not be leachable. Sometimes an ESP of three is too high, but sometimes an ESP of ten or more can be tolerated.

Gypsum Decreases pH of Sodic Soils

Gypsum immediately decreases the pH of sodic soils or near sodic soils from values often over pH 9 but usually over pH 8 to values of from 7.5 to 7.8. These lower values are in the range of acceptability for growth of most crop plants. Probably more than one mechanism is involved. Ca^{++} reacts with bicarbonate $HCO3^-$ to precipitate $CaCO_3$ and release protons (H^+) which decrease the pH. Also, the level of exchangeable Sodium is decreased which lessens the hydrolysis of clay to form hydroxides. These reactions can decrease the incidence of lime and bicarbonate induced iron deficiency.

Gypsum Decreases Bulk Density of Soil

Gypsum-treated soil has a lower bulk density compared with untreated soil. Organic matter can decrease it even more when both are used. Softer soil is easier to till, and crops like it better.

Gypsum Helps Prepare Soil for No-Till Management

A liberal application of gypsum is a good procedure for starting a piece of land into no-till soil management or pasture. Improved soil aggregation and permeability will persist for years and surface-applied fertilizers will more easily penetrate as result of the gypsum.

Gypsum Prevents Crusting of Soil and Aids Seed Emergence

Gypsum can decrease and prevent the crust formation on soil surfaces which result from rain drops or from sprinkler irrigation on unstable soil. Prevention of crust formation means more seed emergence, more rapid seed emergence, and easily a few days sooner to harvest and market. Seed emergence has often been increased by 50 to 100 percent. The prevention of crusting in dispersive soils is a flocculation reaction.

Gypsum Decreases Loss of Fertilizer Nitrogen to the Air

Calcium from gypsum can help decrease volatilization loss of ammonium nitrogen from applications of ammonia, ammonium nitrate, UAN, urea, ammonium sulfate, or any of the ammonium phosphates. Calcium can decrease the effective pH by precipitating carbonates and also by forming a complex Calcium salt with ammonium hydroxide which prevents ammonia loss to the atmosphere. Calcium improves the uptake of nitrogen by plant roots especially when the plants are young.

Gypsum Helps Plants Absorb Plant Nutrients

Calcium, which is supplied in gypsum, is essential to the biochemical mechanisms by which most plant nutrients are absorbed by roots. Without adequate Calcium, uptake mechanisms would fail.

Gypsum Stops Water Runoff and ErosionGypsum improves water infiltration rates into soils and also the hydraulic conductivity of the soil. It provides protection against excess water runoff from especially large storms that are associated with erosion.

Gypsum Decreases Dust Erosion
Use of gypsum can decrease wind and water erosion of soil. Severe dust problems can be decreased, especially when combined with use of water-soluble polymers. Less pesticide and nutrient residues will escape from the surface of land to reach lakes and rivers when appropriate amendments are used to stabilize soil. Gypsum has several environmental values.

Gypsum Improves Soil Structure
Gypsum provides Calcium which is needed to flocculate clays in soil. Flocculation is the process in which many individual small clay particles are bound together to give fewer but larger soil particles. Such flocculation is needed to give favorable soil structure for root growth and air and water movement.

Gypsum Improves Fruit Quality and Prevents Some Plant Diseases
Calcium is nearly always only marginally sufficient and often deficient in developing fruits. Good fruit quality requires an adequate amount of Calcium. Calcium moves very slowly, if at all, from one plant part to another and fruits at the end of the transport system get too little. Calcium must be constantly available to the roots. In very high pH soils, enough Calcium may not be available; gypsum helps make Calcium more available. Gypsum is used for peanuts, which develop below ground, to keep them disease free. Gypsum helps prevent blossom-end rot of watermelon and tomatoes and bitter pit in apples. Gypsum is preferred over lime for potatoes grown in acid soils so that scab may be controlled. Root rot of avocado trees caused by Phytophthora is partially controlled by gypsum.

Gypsum Improves Swelling Clays
Gypsum can decrease the swelling and cracking associated with high levels of exchangeable Sodium on the montmorillonite-type clays. As Sodium is replaced by Calcium on these clays, they swell less and therefore do not easily clog the pore spaces through which air, water and roots move.

Gypsum Makes Slightly Wet Soils Easier To Till
Soils that have been treated with gypsum have a wider range of soil moisture levels where it is safe to till without danger of compaction or deflocculation. This is accompanied with greater ease of tillage and more effective seedbed preparation and weed control. Less energy is needed for the tillage.

Gypsum Prevents Water logging of Soil
Gypsum improves the ability of soil to drain and not become waterlogged due to a combination of high sodium, swelling clay, and excess water. Improvements of infiltration rate and hydraulic conductivity with use of gypsum add to the ability of soils to have adequate drainage.

Gypsum Helps Stabilize Soil Organic Matter
Gypsum is a source of Calcium which is a major mechanism that binds soil organic matter to clay in soil which gives stability to soil aggregates. The value of organic matter applied to soil is increased when it is applied with gypsum.

Gypsum Increases Value of Organic Amendments
Blends of gypsum and organic matter increase the value of each other as soil amendments, especially for improvement of soil structure. Calcium decreases burn out of soil organic matter when soils are cultivated by bridging the organic matter to clay.

Gypsum Corrects Subsoil Acidity
Gypsum can improve some acid soils even beyond what lime (Calcium carbonate) can do for them. Surface crusting can be prevented. The effects of toxic soluble Aluminum can be decreased, even in the subsoil where lime will not penetrate. It is then possible to have deeper rooting with resulting benefits to the crops. The mechanism is more than replacement of acidic hydrogen ions which can be leached from the soil to give higher pH. Hydrogen ions do not migrate rapidly in soils containing clay. It is suggested that the sulfate from gypsum forms a complex (AlS04+) with Aluminum which renders the Aluminum non-toxic. Also suggested is that the sulfate ions react with Iron hydroxides to release hydroxyl ions which give a lime effect to increase soil pH. Gypsum is now being widely used on acid soils.

Gypsum has 17% sulfate, which is the most absorbable form of Sulfur for plants.

Gypsum Makes Water-Soluble Polymer Soil Conditioners More Effective
Gypsum complements or even magnifies the beneficial effects of water-soluble polymers used as amendments to improve soil structure. As is the case with organic matter, Calcium, which gypsum supplies, is the mechanism for binding of the water-soluble polymers to the clay in soils.

Gypsum Makes Magnesium Non-Toxic
In soils having unfavorable Calcium:Magnesium ratios, such as serpentine soils, gypsum can create a more favorable ratio.

Gypsum Improves Water-Use Efficiency
Gypsum increases water-use efficiency of crops. In areas and times of drought, this is extremely important. Improved water infiltration rates, improved hydraulic conductivity of soil, better water storage in the soil all lead to deeper rooting and better water-use efficiency. From 25 to 100 per cent more water is available in gypsum-treated soils.

Gypsum Decreases Heavy-Metal Toxicity: Calcium also acts as a regulator of the balance of the minor and micro-nutrients, such as Iron, Zinc, Manganese and

Copper, in plants. It also regulates non-essential trace elements. Calcium prevents excess uptake of many of them; and once they are in the plant, Calcium keeps them from having adverse effects when their levels get high. Calcium in liberal quantities helps to maintain a healthy balance of nutrients and non nutrients within plants.

Gypsum Decreases the Toxic Effect of NaCl Salinity
Calcium from gypsum has a physiological role in inhibiting the uptake of Na by plants. For species of plants not tolerant to Na, Ca protects from toxicity of Na but not Cl.

Gypsum Keeps Clay From Sticking to Tuber and Root Crops
Gypsum can help keep clay particles from adhering to roots, bulbs and tubers of crops like potato, carrots, garlic and beets.

Gypsum Helps Earthworms to Flourish
A continuous supply of Calcium with organic matter is essential to earthworms that improve soil aeration, improve soil aggregation and mix the soil. Earthworms can do the plowing for no-till agriculture.
See source for references: http://www.usagypsum.com/agricultural-gypsum.aspx

Leaching Excess Sodium and Other Elements from Soils:

(Much of the following is adapted from *Soil Chemistry*, 2nd Edition, by Bohn, McNeal, and O'Connor: Wiley-Interscience 1985, pp255-258. It is recommended as an excellent reference for those who wish to "get serious" about soil chemistry.)

The main requirement in reclaiming salt-affected soils is that sufficient water must pass through the plant root zone to lower the salt concentration to acceptable levels. The passage of 1 meter (39") of leaching water per meter of soil depth under "ponded" conditions normally removes around 80% of the soluble salt from soils. ("Ponding" requires building a dike or dam around the field to be leached, then flooding it with water and keeping it flooded until sufficient water has drained through the soil.)

If the leaching is done under other conditions, such as intermittent ponding or sprinkler irrigation, the quantity of water needed may be lessened quite a lot, perhaps only 1/3 to 1/5 of a meter total. Regardless of the method or amount of water used, it is critical that the leaching water has some place to drain to.

Another and more efficient leaching method is called the basin-furrow method, where the field is nearly leveled and then plowed to leave a series of parallel furrows that meander across the field. Irrigation water is introduced at the high side of the field and allowed to slowly flow back and forth across the field, dissolving salt as it goes.

Whatever method is used, deep ponding with dikes, intermittent ponding, sprinkler irrigation, or the furrow method, soluble Calcium++ ions must be available to replace the excess Sodium+ ions. Usually gypsum, Calcium sulfate, is used for this purpose. The Calcium in the gypsum replaces the Sodium, which binds to the Sulfur in the gypsum, forming Sodium sulfate, and the Sodium sulfate drains away through the subsoil or away from the field if the furrow method is used.

Many soils in arid regions have a layer of Calcium that has accumulated in the subsoil, called a Caliche layer. If a deep-ripping plow is available, the high-Calcium subsoil may be brought to the surface and used for the exchangeable ion. Sulfur will still be needed, at a rate of from 200 to 600 lbs/A or kg/ha, to combine with the Sodium that is released as the Calcium replaces it on the exchange sites. The Sulfur in this case is applied as elemental S, which should be mixed into the brought-up Caliche layer and allowed to sit for a few weeks in order to be oxidized to sulfuric acid and combine with the available Calcium.

Another and rather strange-sounding reclamation method uses high concentrations of salt water from a nearby sea or saline lake. The field is first flooded with full-strength salt water, and then with greater and greater dilutions of salt water, like so:

First flooding/irrigation: Full strength salt water
Second flooding: 1 part salt water, 3 parts fresh water
Third flooding: 1 part salt water, 7 parts fresh water
Fourth flooding: 1 part salt water, 15 parts fresh water
Final flooding: Fresh water only

In all cases, as noted above, the leached salt solution must have somewhere to go, either deeper into the soil or, in the case of the furrow method, to an adjacent area that is not cultivated.

One final and slower method of reclaiming a high-salt soil that does not rely solely on cation exchange, leaching, or deep-ripping is the biological/botanical method. The field is planted to salt-tolerant "weeds" that have strong, deep taproots with the ability to penetrate the subsoil or Caliche layers. As these weeds go through their life cycle, their roots will pierce the hard pan, and then die off, leaving behind organic matter and vertical "drains" for the high salt concentrations to drain away while at the same time bringing up needed minerals from the deep subsoil. For more on this technique, see the book *Weeds: Guardians of the Soil* by Joseph A. Cocannouer, available free at the Journey to Forever website:
http://www.journeytoforever.org/farm_library/weeds/WeedsToC.html#contents

The "Fertile Mulching" Method for Established Orchards, Vineyards, and other Perennials:

Apply the recommended amendment/fertility mix to the area under the plant's canopy, out to and a bit beyond the drip line. Wet the whole area down well, to wash the amendments into the soil. On top of this spread ½ inch of quality compost, and if needed several layers of newsprint or a single layer of corrugated cardboard to keep the weeds and grass down. Wet that down well, then cover the area with 3" of mulch such as straw or ground bark. The feeder roots of the plant will grow up into the newly fertilized zone.

When you wish to apply more fertilizer, rake the top part of the mulch back out of the way, being careful not to damage new feeder roots, apply the new amendments, then rake the mulch back into place.

If rock or clay phosphate have been recommended, it is a good idea to aerate the soil out to the dripline, using a tapered-point digging bar to poke a number of holes about 4" deep. Then spread the phosphate and other amendments and irrigate well before mulching. If time and labor are available, the fertility mix may be poured directly into the holes. This will help get the usually immobile phosphate deeper into the soil. A plug-cutting aerator such as is used for lawns will also work.

Keep the mulch damp during the growing season if possible. This method works great for re-vitalizing aging fruit trees.

hydrogen 1 **H** 1.0079																		helium 2 **He** 4.0026
lithium 3 **Li** 6.941	beryllium 4 **Be** 9.0122												boron 5 **B** 10.811	carbon 6 **C** 12.011	nitrogen 7 **N** 14.007	oxygen 8 **O** 15.999	fluorine 9 **F** 18.998	neon 10 **Ne** 20.180
sodium 11 **Na** 22.990	magnesium 12 **Mg** 24.305												aluminium 13 **Al** 26.982	silicon 14 **Si** 28.086	phosphorus 15 **P** 30.974	sulfur 16 **S** 32.065	chlorine 17 **Cl** 35.453	argon 18 **Ar** 39.948
potassium 19 **K** 39.098	calcium 20 **Ca** 40.078		scandium 21 **Sc** 44.956	titanium 22 **Ti** 47.867	vanadium 23 **V** 50.942	chromium 24 **Cr** 51.996	manganese 25 **Mn** 54.938	iron 26 **Fe** 55.845	cobalt 27 **Co** 58.933	nickel 28 **Ni** 58.693	copper 29 **Cu** 63.546	zinc 30 **Zn** 65.39	gallium 31 **Ga** 69.723	germanium 32 **Ge** 72.61	arsenic 33 **As** 74.922	selenium 34 **Se** 78.96	bromine 35 **Br** 79.904	krypton 36 **Kr** 83.80
rubidium 37 **Rb** 85.468	strontium 38 **Sr** 87.62		yttrium 39 **Y** 88.906	zirconium 40 **Zr** 91.224	niobium 41 **Nb** 92.906	molybdenum 42 **Mo** 95.94	technetium 43 **Tc** [98]	ruthenium 44 **Ru** 101.07	rhodium 45 **Rh** 102.91	palladium 46 **Pd** 106.42	silver 47 **Ag** 107.87	cadmium 48 **Cd** 112.41	indium 49 **In** 114.82	tin 50 **Sn** 118.71	antimony 51 **Sb** 121.76	tellurium 52 **Te** 127.60	iodine 53 **I** 126.90	xenon 54 **Xe** 131.29
caesium 55 **Cs** 132.91	barium 56 **Ba** 137.33	57-70 *	lutetium 71 **Lu** 174.97	hafnium 72 **Hf** 178.49	tantalum 73 **Ta** 180.95	tungsten 74 **W** 183.84	rhenium 75 **Re** 186.21	osmium 76 **Os** 190.23	iridium 77 **Ir** 192.22	platinum 78 **Pt** 195.08	gold 79 **Au** 196.97	mercury 80 **Hg** 200.59	thallium 81 **Tl** 204.38	lead 82 **Pb** 207.2	bismuth 83 **Bi** 208.98	polonium 84 **Po** [209]	astatine 85 **At** [210]	radon 86 **Rn** [222]
francium 87 **Fr** [223]	radium 88 **Ra** [226]	89-102 **	lawrencium 103 **Lr** [262]	rutherfordium 104 **Rf** [261]	dubnium 105 **Db** [262]	seaborgium 106 **Sg** [266]	bohrium 107 **Bh** [264]	hassium 108 **Hs** [269]	meitnerium 109 **Mt** [268]	ununnilium 110 **Uun** [271]	unununium 111 **Uuu** [272]	ununbium 112 **Uub** [277]		ununquadium 114 **Uuq** [289]				

*Lanthanide series	lanthanum 57 **La** 138.91	cerium 58 **Ce** 140.12	praseodymium 59 **Pr** 140.91	neodymium 60 **Nd** 144.24	promethium 61 **Pm** [145]	samarium 62 **Sm** 150.36	europium 63 **Eu** 151.96	gadolinium 64 **Gd** 157.25	terbium 65 **Tb** 158.93	dysprosium 66 **Dy** 162.50	holmium 67 **Ho** 164.93	erbium 68 **Er** 167.26	thulium 69 **Tm** 168.93	ytterbium 70 **Yb** 173.04
**Actinide series	actinium 89 **Ac** [227]	thorium 90 **Th** 232.04	protactinium 91 **Pa** 231.04	uranium 92 **U** 238.03	neptunium 93 **Np** [237]	plutonium 94 **Pu** [244]	americium 95 **Am** [243]	curium 96 **Cm** [247]	berkelium 97 **Bk** [247]	californium 98 **Cf** [251]	einsteinium 99 **Es** [252]	fermium 100 **Fm** [257]	mendelevium 101 **Md** [258]	nobelium 102 **No** [259]

Soil Report 2 Column: Result + Desired

Element		Result	Desired	Comments
Cation Exchange Capacity CEC meq				
pH of Soil Sample				
Organic Matter %				
Anions				
Sulfur S (parts per million ppm)				
Phosphorus as P ppm				
Cations				
Calcium Ca++ ppm	Desired Found Deficit			
Ca Base Saturation 60-70 %				
Magnesium Mg++ ppm	Desired Found Deficit			
Mg Base Saturation 10-20 %				
Potassium K+ ppm	Desired Found Deficit			
K Base Saturation 2-5 %				
Sodium Na+ppm				
Na Base Saturation 1-3%				
Other Bases				
H+ Exch Hydrogen 10-15%				
Minor Elements ppm				
Boron				
Iron Fe				
Manganese Mn				
Copper Cu				
Zinc Zn				
Aluminum				